T0227762

A Practitioner's Guide to

Resampling for Data Analysis, Data Mining, and Modeling

Phillip I. Good

CRC Press
Taylor & Francis Group
Boca Raton   London   New York

CRC Press is an imprint of the
Taylor & Francis Group an **informa** business
A CHAPMAN & HALL BOOK

CRC Press
Taylor & Francis Group
6000 Broken Sound Parkway NW, Suite 300
Boca Raton, FL 33487-2742

© 2012 by Taylor & Francis Group, LLC
CRC Press is an imprint of Taylor & Francis Group, an Informa business

No claim to original U.S. Government works

International Standard Book Number: 978-1-4398-5550-8 (Hardback)

**Visit the Taylor & Francis Web site at
http://www.taylorandfrancis.com**

**and the CRC Press Web site at
http://www.crcpress.com**

Contents

Preface

This text aspires to introduce statistical methodology to a wide audience, simply and intuitively, through resampling from the data at hand.

Practitioners, whether trained as statisticians or having the title *statistician* foisted upon them, soon realize the "normal" distribution almost never occurs in practice, there is no practical equivalent of standard error for measuring the precision of sample percentiles and variances, and it is virtually impossible to explain the meaning of regression coefficients to those whom you are supposed to be helping.

Readers of this text will learn to make use of distribution-free permutation tests, to estimate the precision of sample-based estimates using the bootstrap, and to replace arcane regression methods with the readily interpreted decision tree.

They'll find the resampling methods—permutations, decision trees, and the bootstrap—easy to learn and easy to apply, for these methods require no mathematics beyond introductory high school algebra, yet are applicable to an exceptionally broad range of subject areas, as can be seen by the extensive list of cited applications in Chapter 1.

Practitioners and research workers in the biomedical, engineering, and social sciences, as well as advanced students in biology, business, dentistry, medicine, psychology, public health, and sociology, will find here a practical and easily grasped guide to descriptive statistics, estimation, and testing hypotheses.

Readers of my previous text for CRC, *Applying Statistics in the Courtroom*, will find the resampling methods invaluable at law, as they require the minimum of assumptions and lend themselves readily to interpretation by nonstatisticians.

A hundred or more exercises included at the end of each chapter, plus dozens of thought-provoking questions, will serve the needs of both classroom and self-study. For those whose last statistics course was a long time ago, we've included a glossary of basic concepts in an appendix.

As statistics without software is as useless as Jell-O without a serving dish, we have provided R and Stata code for most resampling methods. We also have made available APL contributed by Valter Sundh, C++, Eviews, Gauss, MATLAB contributed by Melanie Duffin, and SC code for many of the routines at our website: http://statcourse.com/PGsoftware.htm. A list of commercially available software for permutation tests and the bootstrap is provided in Chapter 3, and for decision trees in Section 11.6.

My thanks to Alson Look, Robert Fraczkiewicz, and Aleš Žiberna for their many constructive criticisms, and to Design Science, Cytel Software, FastStone Soft, Rulequest, Stata, and Salford Systems, without whose MathType equation generator, FastStone screen capture utility, See5, StatXact®, Stata®, and CART® statistics packages this text would not have been possible.

I am deeply indebted to my wife, Dorothy, and to the many readers of my other texts who have encouraged me to write this one. If you enter the coupon code VLA2612 when you register for a course at http://statcourse.com, you'll receive a 20% discount.

MATLAB® is a registered trademark of The MathWorks, Inc. For product information, please contact:

The MathWorks, Inc.
3 Apple Hill Drive
Natick, MA 01760-2098 USA
Tel: 508-647-7000
Fax: 508-647-7001
E-mail: info@mathworks.com
Web: www.mathworks.com

Phillip I. Good
Huntington Beach CA
http://statcourse.com
http://zanybooks.com

1

Wide Range of Applications

Due to the lack of high-speed, high-volume computing capability, the theory of resampling methods preceded their practical application by several decades. In 1935, Fisher proposed the use of permutation methods for analyzing categorical data, and in 1937 and 1938, Pitman proposed their use in analyzing continuous measurements. Efron began experimenting with the bootstrap in the late 1970s. Kass (1980) introduced the first decision tree for use in classification by nominal dependent variables.

Resampling Methods

Resampling methods, so called because the data are resampled and reexamined repeatedly to obtain results, consist of permutation tests, the bootstrap, and decision trees. They require no assumptions regarding the frequency distribution of the data—recommending them for courtroom use, yet they can make use of such information if it is available.

In the chapters that follow, you will learn:

- *Permutation tests*, based on repeated relabeling of the data in hand, should be used to do *all* of the following:
  - Make a multivariate comparison of population means
  - Analyze the one-way layout
  - Compare variances
  - Analyze crossover designs
  - Analyze contingency tables

- *Bootstrap methods*, based on repeated resampling from the data in hand, should be used:
  - To obtain approximate confidence intervals for quantiles, standard deviations, and other hard-to-estimate population parameters
  - To validate models
  - To compare means when the population variances are not equal
  - When all other methods fail

- *Decision trees*, based on repeated partitioning and reexamination of the data in hand, can be used for:
  - Classification (assigning subjects to categories)
  - Data mining (helping to focus on the most important variables)
- Presenting regression results in a form readily comprehended by nonstatisticians

Decision trees should be used in preference to regression methods whenever the following is true:

- Predictors are interdependent and their interaction often leads to reinforcing synergistic effects.
- A mixture of continuous and categorical variables, highly skewed data, and large numbers of missing observations add to the complexity of the analysis.

Decision trees are often used in conjunction with bootstrap methods or permutation tests in the analysis of microarrays and other very large data sets with many more variables than independent observations.

Fields of Application

Pitman had to make use of calculators that were cranked by hand. In the 1970s, a colleague of mine complained that if everyone in our company made use of resampling methods our mainframe would be brought to a crashing halt. Today, on an inexpensive desktop computer, results appear almost as soon as the enter key is pressed. No wonder resampling methods are now being applied in virtually every research area, as shown in Table 1.1.

TABLE 1.1

Applications of Resampling Methods

	Bootstrap	Decision Trees	Permutation
Accounting	Biddle et al. 1990	Steadman et al. 2000	
Agriculture		Tittonell et al. 2008	Eden and Yates 1933; Higgins and Noble 1993
Anthropology			Fisher 1936; Gonzalez et al. 2002
Aquatic science	Hoff 2006		Quinn 1987; Ponton and Copp 1997
Archaeology	Ringrose 1992	Reynolds et al. 2008	Klauber 1971; Berry et al. 1980, 1983
Astronomy	Babu and Feigelson 1996; Sen et al. 2009	de la Calleja and Fuentes 2004	Lynden-Bell 1971; Sen et al. 2009; Zucker and Mazeh 2008
Atmospheric science	Fernandez et al. 2005		Adderley 1961; Tukey et al. 1978
Biology	Brey 1990		Daw et al. 1998; Howard 1981
Biotechnology			Vanlier 1996; Park et al. 2002
Botany	Bayer et al. 2009		Mitchell-Olds 1987; Ritland and Ritland 1989
Cardiology		Karaolis et al. 2010	Chapelle et al. 1982
Chemistry	Roy 1994; Jones et al. 1996		vanKeerberghen et al. 1991
Climatology	Robeson 1995		Hisdal et al. 2001
Clinical trials			Howard 1981; Berger 2000
Computer science			Yucesan 1993; Laitenberger et al. 2000
Cytometry		Ladanyi et al. 2004	
Demographics		Balevi 2008	Jorde et al. 1997
Dentistry			Mackert et al. 2001
Diagnostic imaging	Derado 2004		Arndt et al. 1996; Raz et al. 2003; Suckling et al. 2006
Ecology	Adams 1996; Townsend et al. 2006		Pollard et al. 1987
Econometrics	Tambour and Zethraeus 1998; Almasari and Shukur 2003		McQueen 1992; Kennedy 1995
Education			Gliddentracey and Greenwood 1997; Manly 1988

TABLE 1.1 *(Continued)*

Applications of Resampling Methods

	Bootstrap	Decision Trees	Permutation
Endocrinology			O'Sullivan et al. 1989
Environmental science	Park et al. 2002		
Entomology			Bryant 1977; Simmons and Weller 2002
Epidemiology			Glass et al. 1971; Wu et al. 1997
Ergonomics			Valdes-Perez 1995
Finance	Dette et al. 2006	Tseng 2003	
Forensics			Solomon 1986
Genetics	Tivang et al. 1994; Bryan 2004		Karlin and Williams 1984; North et al. 2003
Genetics microarrays	Jiang and Simon 2007		Sohn et al. 2009
Geography		Janssen et al. 1991	Royaltey et al. 1975; Hubert 1978
Geology			Clark 1989; Orlowski et al. 1993
Gerontology			Miller et al. 1997; Dey et al. 2001
Image analysis	Coakley 1996	Homer et al. 2007	Belmonte and Yurgelun-Todd 2001
Immunology	Makinodan et al. 1976		Makinodan et al. 1976; Roper et al. 1998
Law		Arditi and Pulket 2004	Gastwirht 1992; Good 2002
Library science			Dee et al. 1998
Linguistics			Romney et al. 2000
Marketing	Dolnicar and Leisch 2010	Buckinx and Van den Poel 2005	
Medicine	Gong 1986		Feinstein 1973; Tsutakawa and Yang 1974
Mental health service			Tang et al. 2009
Meteorology		Davis and Elder 1994	Gabriel 1979; Tukey 1985
Molecular biology			Barker and Dayhoff 1972
Neurobiology	Cadarso et al. 2006; Landau et al. 2004		Edgington and Bland 1993; Weth et al. 1996
Neurology	Bellac et al. 2008		Lee 2002; Ford et al. 1989
Neuropsycho-pharmacology			Wu et al. 1997
Neuropsychology		Ilgen et al. 2009	Stuart et al. 1997

TABLE 1.1 *(Continued)*

Applications of Resampling Methods

	Bootstrap	Decision Trees	Permutation
Oncology		Kuo et al. 2002	Hoel and Walburg 1972; Spitz et al. 1998
Ophthalmology	Hammoudi et al. 2005		
Ornithology	Lanyon 1987; Senar and Conroy 2004		Cade and Hoffman 1993; Anderson 2006
Paleontology			Marcus 1969; Quinn 1987
Parasitology			Pampoulie and Morand 2002
Pediatrics			Goldberg et al. 1980; Grossman et al. 2000
Pharmacology	Tubert-Bitter et al. 2005	Svetnik et al. 2004	Plackett and Hewlett 1963
Physics			Penninckx et al. 1996
Physiology			Faris and Sainsbury 1990; Zempo et al. 1996
Political science	Mooney 1996		
Psychology			Hollander and Sethuraman 1978; Antretter et al. 2000
Radiology			Raz et al. 2003
Reliability	Choi et al. 1996		Kalbfleisch and Prentice 1980; Nelson 1992
Risk assessment	Bertail and Tressou 2006; Kwon and Moon 2006		
Signal processing	Zoubir and Boashash 1998		
Sociology			Marascuilo and McSweeny 1977; Tsuji 2000
Surgery			Majeed et al. 1996
Taxonomy			Alroy 1994; Fisher 1936
Theology			Witztum et al. 1994
Toxicology			Farrar and Crump 1988, 1991
Virology		Ladanyi et al. 2004	Good 1979
Vocational guidance			Gliddentracey and Parraga 1996; Ryan et al. 1996
Zoology	Sandford and Smith 2002	Kobler and Adamič 1999	Jackson 1990

2

Estimation and the Bootstrap

In this chapter, you'll learn to use the bootstrap to estimate the precision of an estimate, to obtain confidence intervals for population parameters, and to help determine sample size. You'll learn improved bootstrap procedures, including the bias-corrected and -accelerated, blocked, balanced, and adjusted bootstrap. You're provided with computer code in R and several other languages to help you put these resampling methods into practice.

ACCURACY AND PRECISION

Let us suppose Robin Hood and the Sheriff of Nottingham engage in an archery contest. Each is to launch three arrows at a target 50 m (half a soccer pitch) away. The sheriff launches first, and his three arrows land one atop the other in a dazzling display of shooting *precision*. Unfortunately, all three arrows penetrate and fatally wound a cow grazing peacefully in the grass nearby. The sheriff's *accuracy* leaves much to be desired.

Before reading further, be sure you understand which of the two sets of dots in Figure 2.1, one consisting of solid dots and the other of open diamonds, represents accurate shooting (though widely dispersed about the target's center) and which precise.

Precision of an Estimate

We can seldom establish the *accuracy* of an estimate, for example, how close the *sample* median comes to the unknown *population* median. But we may be able to establish its *precision*, that is, how closely estimates derived from successive samples resemble one another.

The straightforward way to establish the precision of the sample median is to take a series of samples from the population, determine the median for each of these samples, and compute the standard deviation of these medians. One can think of at least three reasons why this might not be possible:

1. The method of examination is destructive, as is the case when we need to open a can to check for contamination or burst a condom to measure tensile strength.

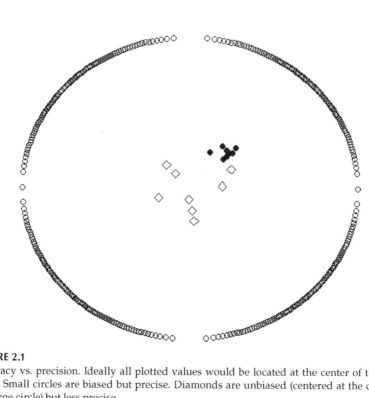

FIGURE 2.1
Accuracy vs. precision. Ideally all plotted values would be located at the center of the large circle. Small circles are biased but precise. Diamonds are unbiased (centered at the center of the large circle) but less precise.

2. The method of examination is prohibitively expensive or time-consuming, or both.

3. The population is hypothetical; for example, when we test a new infant formula, we want to extrapolate our findings to all the children yet to be born.

One practical alternative, known as the *bootstrap*, is to treat the original sample of values as a stand-in for the population and to resample from the original sample repeatedly, with replacement, computing the median each time.

Let's make use of data gathered by students in a sixth-grade class I once taught: their heights as recorded by them in centimeters were: 141, 156.5, 162, 159, 157, 143.5, 154, 158, 140, 142, 150, 148.5, 138.5, 161, 153, 145, 147, 158.5, 160.5, 167.5, 155, 137. Their *median* height is seen to be 153.5 cm.

Suppose we take a bootstrap sample of these data with replacement and with the same number of observations, 22, as were in the original sample. After rearranging this bootstrap sample in increasing order of magnitude for ease in reading, it might look like this:

138.5 138.5 140.0 141.0 141.0 143.5 145.0 147.0 148.5 150.0 153.0 154.0
155.0 156.5 157.0 158.5 159.0 159.0 159.0 160.5 161.0 162

145   (rp50)   157.75

FIGURE 2.2
One-way strip chart of 50 bootstrap medians derived from a sample of heights of 22 students in Dr. Good's sixth-grade class.

Note that several of the values in the original sample have been repeated. This is not unexpected, for when we bootstrap, we sample with replacement. Also, several values in the original sample have not been included. The minimum of this bootstrap sample is 138.5, higher than that of the original sample; the maximum at 162.0 is less; while the median remains unchanged at 153.5.

Let's take another bootstrap sample:

137.0 138.5 138.5 141.0 141.0 142.0 143.5 145.0 145.0 147.0 148.5 148.5 150.0 150.0 153.0 155.0 158.0 158.5 160.5 160.5 161.0 167.5

In this second bootstrap sample, we again find repeated values; this time the minimum, maximum, and median of the sample are 137.0, 167.5, and 148.5, respectively.

As seen in Figure 2.2, the medians of 50 bootstrapped samples drawn from our sample of sixth-graders ranged between 142.25 and 158.25, with a median of 152.75. These bootstrap samples provide a feel for what might have been had we sampled repeatedly from the original population.

To obtain results similar to those shown in Figure 2.2, we may use the following program written in R. It loops repeatedly, drawing a bootstrap sample, computing, and then saving its median.

```
#This program selects 50 bootstrap samples from the
classroom data
#and then produces a boxplot and stripchart of the results.
  class=c(141,156.5,162,159,157,143.5,154,158,140,142,150,
148.5,138.5,161,153,145,147,158.5,160.5,167.5,155,137)
  #Record group sizes
  n = length(class)
  #set number of bootstrap samples
  N =50
  stat = numeric(N) #create a vector in which to store the
results
  #the elements of the vector will be numbered from 1 to N
  #Set up a loop to generate a series of bootstrap samples
  for (i in 1:N){
    #bootstrap sample counterparts to observed samples are
denoted with "B"
    classB= sample (class, n, replace=T)
    stat[i] = median(classB)
```

```
}
boxplot (stat)
stripchart(stat)
```

If you execute this program, you will *not* get the same results I did, because the choice of a bootstrap sample is semirandom based on the computer's internal clock. If you want to get repeatable results, you first have to set a starting number seed for R's random number generator.

```
. set.seed (55)
```

Note: You can replace the 55 by any other integer between 1 and 1,000.

Stata

Once the height data is entered, the following line of code invokes the bootstrap to produce a 95% confidence interval for the *interquartile deviation,*[1] plus a point estimate of its bias.

```
bstrap "summarize height,detail" (r(p75)-r(p25)), reps(100)
nobc nonormal
```

Applying the Bootstrap

By substituting other statistics for the median in the preceding program, one may obtain bootstrap *estimates*[2] of precision for all of the following functionals:

- *Percentiles* of the population's *frequency distribution*
- Central values, such as the *arithmetic mean,* the *geometric mean,* the *median,* and the Hodges-Lehmann estimator
- Measures of dispersion, such as the *variance,* the *standard deviation,* and the *interquartile deviation*
- In an audit, the total expenditure $\Sigma f_i c_i$, where c_i is the cost of the ith item and f_i the frequency with which it occurs

Vickers et al. (2008) used bootstrap methods to obtain confidence intervals for the net benefit of decision curve analysis.

Which Statistic Should We Use?

Most statistics texts read like inferior cookbooks. You want a steak? OK. Here's how you cook it. No choice of recipes. No low-fat, low-salt, low-cal options. But the beauty of the resampling approach is that we are free to use the statistic that is best for the job. Take the seemingly cut-and-dried process of estimating the population mean. Suppose we have a series of independent identically distributed (i.i.d.) observations with cumulative distribution $Pr\{Xi \leq x\} = F[x - \Delta]$ and we want to estimate the location parameter Δ without having to specify the form of the distribution F. If F is the normal or Gaussian distribution, the location parameter Δ corresponds to both the population mean and the population median. And if the loss function is proportional to the square of the estimation error, then the arithmetic mean of the sample is optimal for estimating Δ.

Suppose, on the other hand, that F is symmetric, just as the normal distribution is, so that the population mean is still equal to the population median, but F may include a higher proportion of very large or very small values than a normal distribution. In this case, the sample median is to be preferred as an estimate of Δ.

The standard error of the mean provides an estimate of the precision of the mean. It is widely cited in journal articles. But how are we to estimate and report the precision of the sample median? As we saw in the preceding section, the bootstrap provides the answer.

If we are uncertain whether or not F is symmetric, then our best choice of an estimator is a statistic that was never considered of practical significance in the past, at least until the advent of desktop computers enabled the widespread use of resampling methods such as the bootstrap. It is the Hodges-Lehmann estimator, defined as the median of the $n(n - 1)$ pairwise averages: $\hat{} = median_{i \leq j}(X_j + X_i)/2$.

Table 2.1 itemizes the relative sizes of samples that would be required to obtain equal precision when estimating the location parameter of a normal and a Cauchy distribution, respectively. The latter distribution has a large proportion of relatively large as well as relatively small observations, so that its population mean is infinite; hence, the use of the sample mean as an estimate of the center of the population is inadvisable. On the other hand, the sample median requires many more observations than the sample mean when the data are drawn from a normal distribution. As can be seen from this table, the Hodges-Lehmann estimator represents the ideal compromise between the mean and median. But how are we to estimate and report the precision of this latter estimator? Again, the bootstrap provides the answer, and Exercise 6 in this chapter asks you to apply the bootstrap method described in the "Precision of an Estimate" section to estimate the precision of this estimator.

TABLE 2.1

Relative Sample Sizes

	Distribution	
	Normal	Cauchy
Median	100	100
Mean	64	∞
Hodges-Lehmann	66	100

Confidence Intervals

The problem with single-value (point) estimates is that we will always be in error, unless we can sample the entire population. The solution is an interval estimate or confidence interval where we can have confidence that the true value of the population functional[3] we are attempting to estimate lies between some minimum and some maximum value with a prespecified probability.

For example, to obtain a 90% confidence interval for the variance of sixth-grader's heights, we might exclude 5% of the bootstrap values from each end of Figure 2.3. The result is a confidence interval whose lower bound is 52 cm^2 and whose upper bound is 95 cm^2. Note that our original point estimate of 76.7 cm^2 is neither more nor less likely than any other value in the interval [52, 95].

In the following sections, we use bootstrap confidence intervals to estimate the difference between population means even when the dispersions in the two populations may not be the same, and to test for equivalence.

When Variances Cannot Be Assumed to Be the Same

Student's *t*-test is the standard for comparing two population means based on two sets of measurements obtained from two independent samples drawn from two populations. It relies on the assumption that the variances of the two populations from which we draw our samples are the same. But what if they are not? The bootstrap provides a solution.

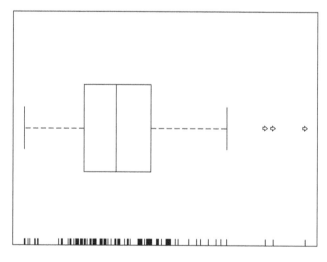

FIGURE 2.3
Box plot and strip chart of variances of 100 bootstrap samples.

Suppose we have *J* observations in our first sample and *K* in our second. Each time we bootstrap, we take *J* observations with replacement from the first sample and *K* from the second, and compute the difference. We repeat several hundred times to form a distribution of these differences. If 0 lies within the 95% confidence interval formed from the 2.5th and the 97.5th percentiles of this distribution, we accept the null hypothesis. Our conclusion is independent of the nature of the distribution from which the observations are drawn.

R

```
#Two Samples
#Efron & Tibshirani, 1993, p. 11
#The observed difference in survival times between treatment
mice
#and control mice is 30.63 days. Determine a 90% confidence
interval
#around this estimate of the true difference using the
percentile bootstrap.
treatmt = c(94, 38, 23, 197, 99, 16, 141) #treatment group
control = c(52, 10, 40, 104, 51, 27, 146, 30, 46) #control group
n = length (treatmt)
m = length (control)
#Find observed difference
mean(treatmt)-mean (control)

#We want to determine whether the difference in means is too
large to have occurred solely by chance if the treatment has
no effect
N = 1000
stat = numeric(N) #create a vector in which to store the results
for (i in 1:N){
#bootstrap sample counterparts to observed samples are denoted
with "B"
treatmtB = sample (treatmt, replace =T)
controlB = sample (control, replace =T)
stat [i] = mean(treatmtB)-mean (controlB)
}
quantile (stat, c(0.05, 0.95))
#If the interval does not include 0, reject the null
hypothesis.
```

Stata

A dummy "treat" variable is used to distinguish the two samples, enabling a stratified bootstrap to be used:

```
drop _all
input treat value
```

```
1 94 1 38 1 23 1 197 1 99 1 16 1 141
0 52 0 10 0 40 0 104 0 51 0 27 0 146 0 30 0 46
end

capture program drop mydiff
program mydiff, rclass
args treat value
sum 'value' if 'treat', meanonly
return scalar mean_treat = r(mean)
sum 'value' if !'treat', meanonly
return scalar mean_contr = r(mean)
return scalar diff = return(mean_treat)-return(mean_contr)
end

set seed 1234
bootstrap "mydiff treat value" r(diff), strata(treat) nowarn
exit

command: mydiff treat value
statistic: _bs_1 = r(diff)

Bootstrap statistics Number of obs = 16
 Number of strata = 2
 Replications = 50
```

Variable	Reps Observed Bias Std. Err. [95% Conf. Interval]
_bs_1	50 30.63492 1.508255 30.15216-29.95811 91.22795 (N)
	-28.19048 85.42857 (P)
	-28.69841 85.12698 (BC)

```
Note: N = normal
 P = percentile
 BC = bias-corrected
```

Testing for Equivalence

Often, we would like to demonstrate that two methods or two treatments are equivalent; that is, their expected values are within d of one another, where the value of d will depend upon the application.

1. Take k observations by each of the methods (that is, $2k$ observations in all).
2. Draw a bootstrap sample consisting of k from the first sample and k from the second sample. Compute the difference in means (or medians) of the two.
3. Draw 400 such bootstrap sample pairs. If 100 − alpha percent of the differences are smaller than d in absolute value, you will have demonstrated equivalence.

To determine the probability of detecting a nonequivalent treatment, assume that the treatment means differ by $D > d$. Add D to all the observations in the first sample and repeat the bootstrap procedure. Your estimate of this probability (the power of the test) is the percentage of differences that are greater than d in absolute value.

Improved Confidence Intervals

Are the confidence intervals we derived in the preceding sections accurate? Is it correct that 90% of the time these intervals will cover the true value of the population parameter? Or is it really a smaller percentage?

Are these intervals efficient? Are they no wider than they should be, so that the probability they include false values of the population parameter is as small as possible?

Unfortunately, the answer for the percentile bootstrap confidence interval in both cases is no. The balance of this section is devoted to providing improved interval estimates.

Bias-Corrected Bootstrap Confidence Interval

Percentile bootstrap intervals are most accurate when the estimate is symmetrically distributed about the true value of the parameter and the tails of the estimate's distribution drop off rapidly to zero, in other words, when the estimate has an almost normal distribution. The "trick" implicit to the bias-corrected BC interval procedure developed by Efron and Tibshirani (1986) is that it behaves as if the sample were drawn from just such a distribution.

Suppose θ is the parameter we are trying to estimate, $\hat{\theta}$ is the estimate, and we were able to come up with a monotone increasing transformation t such that $t(\hat{\theta})$ is normally distributed about $t(\theta)$. We could use this normal distribution to obtain an unbiased confidence interval, and then apply a back-transformation to obtain an almost unbiased confidence interval.

Fortunately, we don't actually have to go through all these steps, but merely agree that we could if we needed to. Though the resulting formula is complicated, it's already incorporated in several computer programs, where it includes a further refinement due to Efron (1987) known as the *bias-corrected and -accelerated* or BC_α *bootstrap*.

One caveat: We do not recommend the use of the bootstrap with samples of fewer than 100 observations. Simulation studies suggest that with small sample sizes, the coverage is far from exact and the endpoints of the intervals vary widely from one set of bootstrap samples to the next. For example, Tu and Zhang (1992) report that with samples of size 50 taken from a normal distribution, the actual coverage of an interval estimate rated at 90% using the BC_α bootstrap is 88%. When the samples are taken from a mixture of two normal distributions (not an uncommon situation with real-life data sets), the actual coverage is 86%. With samples of only 20 in number, the actual coverage is 80%.

More serious when trying to apply the bootstrap is that the endpoints of the resulting interval estimates may vary widely from one set of bootstrap samples to the next. For example, when Tu and Zhang (1992) drew samples of size 50 from a mixture of normal distributions, the average of the limit of 1,000 bootstrap samples taken from each of 1,000 simulated data sets was 0.72, with a standard deviation of 0.16; the average and standard deviation of the upper limit were 1.37 and 0.30, respectively.

Computer Code: The Bias-Corrected and -Accelerated Bootstrap

R

Make sure you are connected to the Internet and then type

```
> install.packages("boot")
```

The installation that includes downloading, unzipping, and integrating the new routines is done automatically. The installation needs to be done once and once only. But each time before you can use any of the boot library routines, you'll need to load the supporting functions into computer memory by typing

```
> library(boot)
```

We'll need to employ two functions from the boot library.

The first of these functions, boot (Data, Rfunction, number), has three principal arguments. Data is the name of the data set you want to analyze, number is the number of bootstrap samples you wish to draw, and Rfunction is the name of an R function you must construct separately to generate and hold the values of existing R statistics functions, such as median or var, whose value you want a bootstrap interval estimate of. For example,

```
> f.median<-function(y,id){
+ median(y[id])
+ }
```

where R knows id will be a vector of form 1:n. Then

```
> boot.ci(boot(classdata, f.median, 400), conf = 0.90)
```

will calculate a 90% confidence interval for the median of the classdata based on 400 simulations.

Note: The first argument in boot() can be a vector or a data frame. The latter would be employed, for example, if one wanted a BCa confidence interval for the difference in means of two samples.

SAS

A macro may be downloaded from http://ftp.sas.com/techsup/download/stat/jackboot.html.

S-Plus

Download the S+Resample package.

```
boot = bootstrap(urdata, median)
boxplot(boot)
plot(boot, 0*boot)

boot = bootstrap(data, median)
limits.bca(boot)
```

Stata

Once the height data are entered, the following line of code invokes the bootstrap to produce a 95% confidence interval for the interquartile deviation, plus a point estimate of its bias.

```
bstrap "summarize height,detail" (r(p75)-r(p25)), reps(100)
nonormal nopercentile
```

Balanced Bootstrap

The purpose of the balanced bootstrap is to ensure that while an observation may be missing from a particular bootstrap sample or be replicated several times, in the total number B of bootstrap samples, each observation will appear the same number B of times. The result is a slight decrease in the variance of the bootstrap and a reduction in computation time. This reduction is not of practical significance in most applications, with the exception of those of high dimensionality, such as cluster analysis and correspondence analysis.

To obtain a balanced bootstrap, fill a vector with the observations in the original sample replicated B times. Randomly permute the observations in the vector, then take successive bootstrap samples starting at one end of the permuted vector, with the final bootstrap sample being taken at the opposite end.

Tilted Bootstrap

The tilted bootstrap is of particular value with estimates such as the sample variance and the 90th percentile, which are strongly dependent upon outlying values. Instead of giving equal probability to each of the observations in the original sample, greater weight is given to some of the observations and less weight to others. This technique is also known as importance sampling.

Suppose we are testing the hypothesis that the population mean is 0 and our original sample contains the observations –2, –1, –0.5, 0, 2, 3, 3.5, 4, 7, 8. Bootstrap samples containing 7 and 8 are much more likely to have large means, so instead of drawing bootstrap samples such that every observation has the same probability 1/10 of being included, we weight the sample so that larger values are more likely to be drawn, selecting –2 with probability 1/55 for example, –1 with probability 2/55, and so forth, with the probability of selecting 8 being 9/55th. Let $I(S^* > 0) = 1$ if the mean of the bootstrap sample is greater than 0 and 0 otherwise. The standard estimate of the proportion of bootstrap samples greater than zero is given by the formula

$$\frac{1}{B} \sum_{b=1}^{B} I(S_b^* > 0)$$

where B is the number of bootstrap samples. Since the elements of our bootstrap samples were selected with unequal probability, we have to use a weighted estimate of the proportion:

$$\frac{1}{B} \sum_{b=1}^{B} I[S_b^* > 0] \prod_{i=1}^{n} \pi_i^{n_{ib}} n^n$$

where πi is the probability of selecting the ith element of the original sample of size n and nib is the number of times the ith element appears in the bth bootstrap sample of size n. Efron and Tibshirani (1993, p. 355) found a sevenfold reduction in variance using importance sampling with samples of size 10.

Tilted bootstrap methods also have been proposed to reduce the effects of outliers in the bootstrap resamples (see, for example, Davison and Hinkley, 1997, p. 166).

Block Bootstrap

A glance at a sixth-grade classroom will reveal, much to the dismay of the sixth-grade boys, that girls are taller than boys. To construct bootstrap confidence intervals in situations where the effects of age, sex, concurrent medications, and other factors may well overwhelm that of the object of the investigation, separate bootstrap samples should be drawn from each of the factor combinations.

Statistical tests applied to EEGs and MRIs have to deal with the uncertainty related to time series, which cannot be regarded as independent and identically distributed samples from a given process. The *circular block bootstrap* should be employed in order to respect the temporal dependencies of the data. This consists of drawing blocks of the time series rather than independent observations. The block length needs to be adapted to the range of temporal dependencies and the number of volumes. For adequate block lengths, this method preserves spatial correlation, and formally leads to consistent confidence intervals of such correlations (Lahiri, 2003).

Iterated Bootstrap

We want confidence intervals that are:

1. More likely to contain the correct value of the functional being estimated than any specific incorrect value
2. Less likely to contain incorrect values

We can reduce the width of a confidence interval (and thus the likelihood that it will contain incorrect values) by lowering the degree of required confidence: a 90% confidence interval is narrower than a 95% confidence interval, and an 80% confidence interval is narrower still.

We can also narrow the width of a confidence interval by increasing the size of the original sample. A 95% confidence interval for the mean of a normal distribution is given by the formula $\bar{X} \pm 1.96s/\sqrt{n}$, where s is the standard deviation of the sample. If you want to halve the size of such a confidence interval, you have to increase the sample size from 100 to 400. Such confidence intervals, whose width for large samples is proportional to $n^{-\frac{1}{2}}$, are said to be *first-order exact*.

The percentile bootstrap confidence interval is also first-order exact. BC_α intervals are second-order exact in that their width for large samples is proportional to n^{-1}. That is, to halve the size of a confidence interval, you would only have to double the sample size.

Bootstrap iteration can improve the accuracy of bootstrap confidence intervals; that is, it can make it more likely that they will cover the true value of the population functional, and it can also increase the rate at which the width decreases from second to third order (proportional to $n^{3/2}$).

Let I be a $1 - \alpha$ level confidence interval for θ whose actual coverage $\pi_\theta[\alpha]$, depends upon both the true value of θ and the hoped for coverage, $1 - \alpha$. In most cases $\pi_\theta[\alpha]$ will be larger than $1 - \alpha$. Let α' be the value of α for which the corresponding confidence interval I' has probability $1 - \alpha$ of covering the true parameter. We can't solve for α' and I' directly. But we can obtain a somewhat more accurate confidence interval I^* based on α^*, the estimate of α' obtained by replacing θ with its plug-in estimate and the original sample with a bootstrap sample. Details of the procedure are given in Martin (1990, pp. 113–114). The iterated bootstrap, while straightforward, is computationally intensive. Suppose the original sample has 30 observations, and we take 300 bootstrap samples from each of 300 bootstrap samples. That's 30 × 300 × 300 = 2,700,000 values, plus an equal number of calculations![4]

When the Form of the Distribution is Known

If we know the form of the population distribution (exponential, Weibull), a Monte Carlo simulation can provide more accurate confidence intervals than any of the nonparametric bootstraps. For example, if we know the observations come from a gamma distribution, we can use the mean and variance of the sample to estimate its shape and scale parameters. We would then draw a series of random samples from a gamma distribution with these parameters and use these samples to obtain a confidence interval for the population characteristic of interest. When I attempted to derive an 80% confidence interval for the 30th percentile of a gamma distribution with shape = 3.5 and scale =2 (a chi-square distribution with 7 degrees of freedom), the coverage of the nonparametric bootstrap confidence interval was 78%, while making use of the knowledge that the distribution was of gamma form increased the coverage to 80%. (Similar results are obtained with other distributions and other population functionals.)

Schall (1995) used the parametric bootstrap to assess bioequivalence using parameter estimates obtained from an initial analysis of variance. Proving that one can't make a silk purse out of a sow's ear, Kwon and Moon (2006) make a series of rash assumptions about the parametric form of the extreme tail of a distribution, then use the parametric bootstrap to assess the risk of a dam overflowing.

Estimating Bias

When an estimate is inaccurate or biased, we would like an estimate of its bias, that is, the amount by which its expected value differs from the quantity to be estimated. The bootstrap can also help us here. Recall that while the variance of our sample of sixth-graders' heights is 76.7 cm^2, the mean of the variances of 100 bootstrap samples drawn from is 71.4 cm^2. Thus, our original estimate of the population variance would appear to be biased upward by 5.3 cm^2.

More generally, let $E(X)$ denote the expected or mean value of a random variable X. An estimate $\theta[X]$ based on a sample is also a random variable; we define the bias of $\theta[X]$ as $b = E(\theta[X] - \theta)$, where θ is the population parameter we are trying to estimate. A bootstrap estimate for the bias b of $\theta[X]$ is given by $b* = \Sigma_i (\theta_i^* - \theta[X])/k$, where θ_i^* is the ith bootstrap sample estimate of θ for $1 \le i \le k$.

An Example

A small-scale clinical trial was conducted to show the FDA that a product produced at a new plant was equivalent to the product produced at the old plant. In this crossover trial, eight patients received in random order each of the following:

- A patch containing hormone that was manufactured at the old site
- A patch containing hormone that was manufactured at the new site
- A patch without hormone (placebo) that was manufactured at the new site

To establish equivalence, the U.S. Food and Drug Administration (FDA) required that the absolute value of the ratio of θ to μ be less than or equal to 0.20, where θ was the expected value of the improvement at the new site with respect to the old, $E(\text{new}) - E(\text{old})$, and μ was the expected value of the improvement at the old site with respect to the placebo, $E(\text{old}) - E(\text{placebo})$.

The natural estimate for θ is the average of the old-patch hormone level in patients minus the average of the new-patch hormone level in patients. Similarly, the natural estimate of μ is the average of the new-patch hormone level in patients minus the average of the placebo hormone level in patients. The plug-in estimate, which is the ratio of these two estimates, is the natural estimate of the ratio.

The absolute value of the plug-in estimate for θ/μ for the data in Table 2.2. is 0.07. This is considerably less than the FDA's criterion of 0.20. Unfortunately, our estimate is biased, both because it is a ratio and because the same factor $E(\text{old})$ appears in both the numerator and the denominator. We may use the bootstrap to estimate the bias. Efron and Tibshirani (1993, Chapter 10) generated 400 bootstrap samples and found the bootstrap estimate of bias to be only 0.0043.

Applying this bias adjustment, we see that our estimate is still considerably less than 0.20 in absolute value. Warning: The bootstrap is unreliable for small samples. More than 100 patients would be needed to get a reliable estimate of the bias in the present case.

TABLE 2.2

Estimates Used to Determine Bias

									Average
Denominator	8,406	2,342	8,187	8,459	4,795	3,516	4,796	10,238	6,342
Numerator	−1,200	2,601	−2,705	1,982	−1,290	351	−638	−2,719	−452

Determining Sample Size

Original Sample

In planning an investigation we need to balance two considerations:

1. Controlling costs by keeping the sample size as small as possible
2. Ensuring that the sample is sufficiently large that the confidence interval we derive will both cover the true parameter value with the desired probability and yet be as narrow as possible so as to exclude false values

If we know that our observations will have some well-tabulated parametric distribution such as the normal, the exponential, or the binomial, then our task is a relatively simple one, for there is plenty of software at hand to assist in doing the necessary calculations. If we already have some data—enough to convince us that the observations do not come from some well-tabulated distribution—then we may use the bootstrap to estimate the necessary sample size.

The process is an iterative one:

We obtain an initial estimate for the sample size by assuming that the observations come from some well-tabulated distribution.

Next, we take repeated bootstrap samples from the data we have at hand to obtain an estimate of the desired α level cutoff value for the parameter of interest. For example, $\alpha = 5\%$. (Note that the statistical method we apply to each bootstrap sample could be either a parametric or a resampling method.)

Let Δ denote the largest error in our estimate we are willing to tolerate. Add this value to each of the bootstrap estimates we obtained at step 2. Let β denote the percentage of these estimates that exceed the α level cutoff value obtained at step 2.

We would like β to be as large as possible, typically $\beta \geq 80\%$. If $\beta < 80\%$, we increase the sample size and repeat steps 2 and 3. If $\beta > 80\%$, we would decrease the sample size and repeat steps 2 and 3. To obtain the optimal sample size, we would end the iterative process only when $\beta = 80\%$ (or is sufficiently close to this value).

Suppose that a small-scale preliminary study yielded the following results for survival with and without treatment:

Control = c(52, 10, 40, 104, 51, 27, 146, 30, 46)
Treatment = c(94, 38, 23, 197, 99, 16, 141)

As survival tends to follow an exponential distribution, we would take the logarithms of the observations before proceeding further:

```
n = 15 #guess as to necessary sample size
NO = 1000 #number of repetitions of outer loop
NI= 1000 #number of repetitions of inner loop
stat = numeric(N1) #create a vector in which to store the
results
cnt=0
for (k in 1:NO){
  for (i in 1:NI){
    #bootstrap sample counterparts to observed samples are
denoted with "B"
    #bootstrap sample may be much larger than original sample
    treatmtB = sample (Ltreatmt, n, replace =T)
    controlB = sample (Lcontrol, n, replace =T)
    stat [i] = mean(treatmtB)-mean (controlB)
  }
  Q= quantile (stat, c(0.05, 0.95))
  If (0<Q[[1]] or 0 > Q[[2]]) cnt=cnt+1
}
power = cnt/NO
```

Bootstrap Sample

Should each bootstrap sample have the same number of observations as the original sample? The answer to this question depends on the use to which the bootstrap samples will be put. If your objective is to use the original sample to estimate the size a sample from the same population ought be to obtain a predetermined power level (a topic we discuss at length in the next chapter), then you will need to experiment with a variety of sample sizes, some smaller, some larger than the original sample. Bootstrap sample sizes less than the size of the original sample are recommended when the original sample size is extremely large, as would be the case when analyzing microarrays and consistent estimates are desired (Lee and Pun, 2006).

But if your objective is to estimate the precision of the original estimates, then your bootstrap samples need be precisely the same size as the original sample. Here is why: Recall that the precision of a sample mean is proportional to the square root of the number of observations in the sample. Smaller samples yield less precise estimates, while estimates based on larger samples are more precise.

How many bootstrap samples should one take? For that matter, how large ought the original sample be for application of the bootstrap to make practical sense?

The answers to these two questions are interdependent. The quickest way to spot that the original sample size is inadequate is to take two sets of 100 bootstrap samples and compare the results. If the results are radically different, this means that the original sample lacked sufficient information concerning the estimator in question. Taking 10,000 bootstrap samples as Chernick (2008) suggests is like putting pancake makeup on a zit; it merely conceals, not eliminates, the problem.

The good news is that bootstrap estimates of a population's location parameter using the improved methods described in previous sections tend to be fairly precise even when the original sample is small; the bad news is that estimates that depend on the tail of a distribution, such as dispersions, bias, and the 90th and 95th percentiles, require quite large samples in order to be estimated with sufficient precision.

As we show in Chapter 10, the bootstrap also can be used to determine if the sample used to develop a model was sufficiently large.

Summary

In this chapter, you learned and quickly mastered the use of the primitive or percentile bootstrap to obtain estimates of the precision of estimates. You were shown how the bootstrap could be used to compare population means even when the population variances were not the same. And you were shown how bootstrap confidence intervals could be used to test for equivalence.

Unfortunately, bootstrap confidence intervals are neither exact nor unbiased; that is, they are as likely to cover false values of the unknown parameter as they are to cover the true one. Consequently, you were given a variety of methods for improving the accuracy of bootstrap confidence intervals, including the BC_α bootstrap, the balanced bootstrap, the tilted bootstrap, and the iterative bootstrap. You were provided with computer code to aid you in putting these methods into practice.

To Learn More

The bootstrap has its origins in the seminal work of Jones (1956), McCarthy (1969), Hartigan (1969, 1971), Simon (1969), and Efron (1979, 1982). Among its earliest applications to real-world data are Makinodan et al. (1976) and Gong (1986). For further examples of the wide applicability of the bootstrap method, see Chapter 1 as well as Chernick (2008).

Bias-corrected and -accelerated bootstrap confidence intervals and a computationally rapid approximation known as the ABC method are described in Efron and Tibshirani (1993, Chapter 14). For deriving improved two-sided confidence intervals, see the discussion by Loh following Hall (1988, pp. 972–976). See also DiCiccio and Romano (1988) and Tibshirani (1988).

The moving blocks algorithm was introduced by Liu and Singh (1992). The balanced bootstrap is employed for a study of the stability of nonlinear principal components by Linting et al. (2007) and for the correspondence analysis of microarray time-course data by Tan et al. (2004). Algorithms for balanced bootstrap simulations are provided by Gleason (1988). Importance sampling for bootstrap tail probabilities is discussed in Johns (1988), Hinkley and Shi (1989), Do and Hall (1991), and Phipps (1997).

Hall (1986), Beran (1987), Hall and Martin (1988), and Hall (1992) describe iterative methods, known respectively as bootstrap pivoting and bootstrap inverting, that provide third-order accurate confidence intervals. Loh (1987, 1991) describes a bootstrap calibration method that yields confidence intervals that in some circumstances are fourth-order accurate.

For guidance on the number of bootstrap replications, see Andrews and Buchinsky (2000).

In Chapter 10, we consider another major application of the bootstrap: model validation along with its application to the analysis of microarrays, EEGs, and satellite imagery.

Potential flaws in the bootstrap approach are considered by Schenker (1985), Wu (1986), DiCiccio and Romano (1988), Efron (1988, 1992), Knight (1989), Gine and Zinn (1989), and Molinaro et al. (2005). Canty et al. (2000) provide a set of diagnostics for detecting and dealing with potential error sources. Good and Hardin (2009, p. 102) list potential limitations to the bootstrap approach.

Exercises[5]

1. Use 50–100 bootstrap samples to obtain 80% confidence intervals for the standard deviation and median of the heights of my sample of 22 sixth-graders. How do these values compare with the standard deviation and the median of the original sample?

2. If every student in my sixth-grade class grew 6 cm during the school year, what would the mean, median, and variance of their new heights have been?

   If their heights had been measured in inches rather than centimeters, what would the mean, median, and variance of their heights have been? (Assume that 2.54 cm = 1 in.) (Hint: Step away from your computer—you won't need it—and think about your answer.)

3. A possible problem has occurred in the manufacturing plant and management is concerned that the failure rate has changed for the worse. Is the current process the same as the historical process? For

all practical purposes, the processes will be deemed equivalent if the means are within 5 units of each other.

> Historical = c(28, 27, 26, 32, 36, 37, 27, 35, 40, 34, 43, 33, 33, 31, 30, 31, 34, 33, 35, 26, 32, 31, 23, 17, 20, 20, 19, 26, 31, 24, 35, 36, 33, 20, 33, 35, 35, 22, 33, 37, 38, 29, 34, 29, 28, 32, 36, 41, 41, 33, 23, 39, 28, 26, 25, 26, 27, 37, 34, 33, 33, 31, 29, 20, 19, 34, 19, 20, 30, 52, 51, 42)

> Current = c(45, 31, 41, 27, 26, 42, 35, 42, 29, 30, 28, 25, 29, 25, 22, 25, 34, 31, 32, 23, 26, 33, 31, 28, 34, 32, 31, 34, 28, 32)

4. The audit: Recently, 30 invoices were selected at random from 10,000. Sixteen of these invoices were found to contain errors in the following amounts in dollars: 47.8, 50.41, 38.73, 62.26, 20.50, 56.54, 53.67, 26.42, 44.10, 71.36, 51.3, 70.62, 56.54, 33.37, 26.79, 79.50. Use the bootstrap to derive a lower confidence bound for the money that was overbilled.

5. Suppose you wanted to estimate the population variance. Most people would recommend you divide the sum of squares about the sample mean by the sample size minus 1, while others would say that is too much work—just the divide the sum of squares about the sample mean by the sample size. What do you suppose the difference is? (Need help deciding? Try this experiment: Take successive bootstrap samples of size 11 from the sixth-grade height data and use them to estimate the variance of the entire class.)

6. "Use the sample median to estimate the population median and the sample mean to estimate the population mean." Better still, use the Hodges-Lehmann estimator. This sounds like good advice, but is it? Use the technique described in the previous exercise to check out which of the three approaches is most desirable.

7. In answering the preceding question, did you use the same size for the bootstrap sample from each of the four sets of observations? Would there be an advantage to taking a bootstrap sample that was larger or smaller than the original sample? Why or why not?

Endnotes

1 See Appendix A for a glossary of statistical terminology.
2 If you are already familiar with the terms in *italics*, read on. But for the benefit of those for whom it's been a long time since their last statistics class, we define these terms and comment on their applications in Appendix A.
3 See the Glossary for a definition of this and all other italicized terms whose definitions are not provided in the surrounding text.
4 A third iteration involving some 800 million numbers will further improve the interval's accuracy.
5 Answers to selected exercises will be found online.

3

Software for Use with the Bootstrap and Permutation Tests

In the absence of high-speed computers and the supporting software, the resampling methods, indeed almost all statistical procedures, would merely be interesting theoretical exercises. Though introduced in the 1930s, the numerous, albeit straightforward, calculations required by the resampling methods—resampling with replacement for the bootstrap, cross-validation for decision trees, and rearrangements for permutation tests—were beyond the capabilities of the primitive calculators then in use. They were soon displaced by less powerful, less accurate parametric approximations that made use of tables. Today, with a powerful computer on every desktop, resampling methods have resumed their dominant role and table lookup is an anachronism.

But not without the software! We have two alternative approaches: One can program it oneself in a computer language such as APL, C++, MATLAB, R, or SAS macros, or make use of a menu-driven program such as Microsoft Excel, S-Plus, Stata, StatXact, or Testimate. One might well make use of both types of software: menu-driven programs because of their convenience and a programming language because sooner or later you'll encounter a new application beyond a menu-driven program's capabilities. I make use of all of these programs in my work as a statistical consultant.

AFNI

ANFI incorporates permutation tests for use in analyzing neuroimages. Versions for Macs, PCs, and Unix computers may be downloaded from http://afni.nimh.nih.gov/afni/download/afni.

Blossom Statistical Analysis Package

Blossom is a free interactive program utilizing multiresponse permutation procedures (MRPPs) for grouped data, agreement of model predictions, circular distributions, goodness of fit, least absolute deviation, and quantile regression.

An online manual in HTML format contains many worked through examples. It runs under Windows. See http://www.fort.usgs.gov/products/software/blossom/blossom.asp.

Eviews

The makers of this program designed for the analysis of economic and other time series may not be able to tell a permutation test from a bootstrap, but if your objective is to test series components and residuals for independence, then this is the program to buy. It includes routines to @permute and @resample from matricies. Order a 30-day trial version for use with Windows from http://www.eviews.com/.

HaploView

This special purpose program designed for use with microarrays provides for permutation testing for association significance. It may be downloaded without charge from http://www.broadinstitute.org/mpg/haploview/index.php.

MATLAB®

MATLAB can be used to program your own bootstrap and permutation routines. It runs on PCs, Macs, and Unix-based computers. For more information consult http://www.mathworks.com/products/index.html?ref=fp.

NCSS

This software generates a permutation distribution for Hotelling's T^2. Download a free trial from http://www.ncss.com/download_freetrial.html.

PAUP

For phylogenetic analysis, PAUP incorporates both bootstrap and permutation procedures. With it one can analyze molecular sequences or morphologic data. It runs on Macs, PCs, and Unix-based computers. Download from http://paup.csit.fsu.edu/.

R

R is a do-it-yourself programming language specifically designed for use by statisticians. This means that functions like mean(), quantiles(), binom(), glm(), plot(), sample(), and tree() are precompiled and ready for immediate use. Moreover, it is easy to add your own statistical functions. You can download R for free from http://www.cran.r-project.org/ and obtain manuals and directions to self-help bulletin boards from the same Internet address. Many user-written precompiled libraries of resampling functions are available for download. For example, permtest compares two groups of high-dimensional signal vectors derived from microarrays, for a difference in location or variability.

R runs on PCs, Macs, and Unix-based computers. The following courses get you started with R and then apply it to resampling methods:

http://statcourse.com/intro2R.htm

http://statcourse.com/application.htm

SAS

SAS is a highly expensive menu-driven program best known for its ability to build customized tables and graphs. Widely available among corporate users, resampling methods can be added via a macro language. The cumbersome nature of this language, a giant step backward from the C++, in which SAS is written, makes it extremely difficult and time-consuming to write and debug your own resampling methods. SAS Proc MultiTest is recommended for obtaining accurate significance levels for multiple tests.

S-Plus

S-Plus is a menu-driven variant of R aimed at corporate users. It's expensive but includes more built-in functions, the ability to design and analyze sequential trials, a huge library of resampling methods, plus a great deal of technical support. It works with PCs and UNIX-based computers. Check it out at http://spotfire.tibco.com/Products/SPLUS-Client.aspx.

SPSS Exact Tests

This is an add-on module for SPSS. Limited to one-, two-, and k-sample (square deviate) permutation tests. Check it out at http://www.spss.com/media/collateral/statistics/exact-tests.pdf.

Stata

Stata is a comprehensive menu-driven program that allows you to generate bootstrap confidence intervals for the results of virtually any statistical procedure from population means to linear, logistic, and quantile regression coefficients. The latest version allows you to perform permutation tests for virtually all situations, including, unfortunately, some tests, such as multifactor analysis and multiple regression, for which permutation methods are not recommended. See http://stata.com.

Statistical Calculator

Like R, the Statistical Calculator is an extensible statistical environment, supplied with over 1,200 built-in (compiled C) and external (written in SC's C-like language) routines. It includes permutation-based methods for contingency tables (chi-square, likelihood, Kendall S, Theil U, kappa, tau, odds ratio), one- and two-sample inferences for both means and variances, correlation, and multivariate analysis (MV runs test, Boyett/Schuster, Hotelling's T^2). It also includes ready-made bootstrap routines for testing homoscedacity, detecting multimodality, plus general bootstrapping and jackknifing facilities. See http://www.mole-soft.demon.co.uk/.

StatXact

Menu-driven StatXact is the package to purchase for the analysis of contingency tables with categorical or ordered data. It includes power and sample size calculations. See www.cytel.com.

Testimate

Testimate includes effect size measures and confidence intervals, tests for difference and noninferiority, the Su-Wei test for difference or quotient of medians, the Wei-Lachin procedure, and the Cochran-Armitage test. It is for PCs only. See http://www.idvgauting.com/cms/index.php?id=testimate&L=1.

4

Comparing Two Populations

In this chapter, we introduce the permutation test and apply it to the comparison of two populations on the basis of two samples, one from each population. Examples of such comparisons include the following:

Will a new medication reduce blood pressure with fewer side effects than the current most popular treatment?

Is our manufacturing process still under control, or are the units we produce more variable than before?

Use of the permutation distribution guarantees that our significance levels will be exact, not approximations, and that the samples needed to detect differences between populations will be kept as small as possible.

A Distribution-Free Test

My next-door neighbor, an engineer, constantly complains (justifiably for the most part) about being put upon at work. It seems they've made him their designated statistician despite his lack of training or interest in the subject.

"My boss brings me six numbers, three from each of two samples, and tells me to run a t-test. Can I do that?"

"Are the data normally distributed?"

"I wish."

"If you had at least six observations in each sample, and could be sure the data were almost normal, you could probably use a t-test. But there is another, better test, every bit as powerful as Student's t, that is, distribution-free."

Here are the numbers he gave me: 121, 118, 110, 34, 12, 22. The first three counts were from heat-treated wires, and the last three were from untreated. The difference was clearly significant.

The null hypothesis is that the heat treatment had no effect. The alternative of interest was that preheating would increase tensile strength.

Under the null hypothesis, the labels "treated" and "untreated" provide no information about the outcomes: the observations are expected to have more or less the same values in each of the two experimental groups. If they do differ, it should only be as a result of some uncontrollable random fluctuation. Thus, if this null or no difference hypothesis were true, we are free to exchange the labels.

The next step in the permutation method is to choose a test statistic that best discriminates between the hypothesis and the alternative. The statistic I chose was the sum of the counts in the treatment group. If the alternative hypothesis is true, most of the time this sum ought to be larger than the sum of the counts in the untreated group. If the null hypothesis is true, that is, if it doesn't make any difference which treatment the wires receive, then the sums of the two groups of observations should be approximately the same. One sum might be smaller or larger than the other by chance, but most of the time the two shouldn't be all that different.

The third step in the permutation method is to compute the test statistic for each of the possible relabelings and compare these values with the value of the test statistic as the data were labeled originally. The value of the test statistic for the observations as originally labeled is 349 = 121 + 118 + 110.

I began to rearrange (permute) the observations, randomly reassigning the six labels, three "treated" and three "untreated," to the six observations. For example, treated, 121 118 34, and untreated, 110 12 22. In this particular rearrangement, the sum of the observations in the first (treated) group is 273. I repeated this step till all $\binom{6}{3}$ = 20 distinct rearrangements had been examined.

	First Group	Second Group	Sum of First Group
1.	121 118 110	34 22 12	349
2.	121 118 34	110 22 12	273
3.	121 110 34	118 22 12	265
4.	118 110 34	121 22 12	262
5.	121 118 22	110 34 12	261
6.	121 110 22	118 34 12	253
7.	121 118 12	110 34 22	251
8.	118 110 22	121 34 12	250
9.	121 110 12	118 34 22	243
10.	118 110 12	121 34 22	240
11.	121 34 22	118 110 12	177
12.	118 34 22	121 110 12	174
13.	121 34 12	118 110 22	167
14.	110 34 22	121 118 12	166
15.	118 34 12	121 110 22	164
16.	110 34 12	121 118 22	156
17.	121 22 12	118 110 34	155
18.	118 22 12	121 110 34	152
19.	110 22 12	121 118 34	144
20.	34 22 12	121 118 110	68

The sum of the observations in the original treated sample, 349, is equaled only once and never exceeded in the 20 distinct random relabelings. If chance alone is operating, then such an extreme value is a rare, only 1-time-in-20 event. If I reject the null hypothesis and embrace the alternative that the treatment is effective and responsible for the observed difference, I only risk making an error and rejecting a true hypothesis 1 in every 20 times.

A Little Math

Why just 20 possibilities? And why did we use just the sum of the observations in the first sample as our test statistic instead of the difference in means or, for that matter, Student's *t*?

We can choose three out of six observations to be given the "treatment" label in $6 \times 5 \times 4$ ways, or $6!/3!$. But $3 \times 2 \times 1 = 3!$ of each of these ways is just a reordering of the three we chose. So we have exactly $6 \times 5 \times 4/(3 \times 2)$ distinct rearrangements. Note that the "untreated" labels are automatically assigned to the three observations that are left over; no choice here.

If we had samples of unequal size, say four in one group and three in the other, we would have had 7 choose 3, or $7!/\{3!(7-3)!\}$ ways to label them.

Recall that the *t*-statistic has the formula

$$\frac{\sum_{j=1}^{n} X_j/n - \sum_{j=n+1}^{n+m} X_j/m}{\sqrt{\sum_{j=1}^{n} (X_j - \sum_{j=1}^{n} X_j/n)^2 + \sum_{j=n+1}^{n+m} (X_j - \sum_{j=n+1}^{n+m} X_j/m)^2}} \sqrt{n+m-2}$$

Note that the numerator can be rewritten as

$$\left(\frac{m+n}{mn}\right) \sum_{j=1}^{n} X_j - \sum_{j=1}^{n+m} X_j/m$$

Eliminating all the terms remaining unchanged (*invariant*) under permutations, including the sample sizes, the sum of all the observations, and the sums of the squares of all the observations, we are left with just the sum of the observations in the first sample. In the balance of this text, we will simplify our computations as often as we can by removing invariants.

Some Statistical Considerations

The preceding experiment raises a number of concerns that are specifically statistical in nature. How do we go about framing a hypothesis? If we reject a hypothesis, what are the alternatives? Why must our conclusions always be in terms of probabilities rather than certainties? We consider each of these issues in turn.

Framing the Hypothesis

The hypothesis of no treatment effect that we tested in this example is what statisticians term a *null hypothesis*. It is based on the assumption that the samples we test are all drawn from the same hypothetical population, or at least from populations for which the variable we observe has the same distribution. Thus, the null hypothesis that the distribution of the heights of sixth-grade boys is the same as the distribution of the heights of sixth-grade girls clearly is false, while the null hypothesis that distribution of the heights of sixth-grade boys in Orange County is the same as the distribution of the heights of sixth-grade boys in neighboring Los Angeles County may well be true.

The majority of hypotheses considered in this text are null hypotheses, but they are not the only possible hypotheses. For example, we might want to test whether heat-treated wires would show an increase of at least 10 units in tensile strength. As it is difficult to test such a hypothesis, we normally would perform an initial transformation of the data, subtracting 10 from each of the observations in the treated sample. Then, we would test the null hypothesis that there is no difference between the observations from the original untreated sample and the transformed observations.*

Hypothesis vs. Alternative

Whenever we reject a hypothesis, we accept an alternative.† In the present example, when we rejected the null hypothesis, we accepted the alternative that heat had a beneficial effect. Such a test is termed *one-sided*. A two-sided test would also guard against the possibility that heat had a detrimental effect. A two-sided test would reject for both extremely large and extremely small values of our test statistic. See the "Matched Pairs" section for a more extensive discussion. In more complex situations, such as those considered in Chapter 7, many different tests are possible, depending on which alternatives are of greatest interest.

In reviewing the possible outcomes of heat treatment, the phrase "most of the time" was used repeatedly. Yet something that is predominantly true may still be false on occasion. An ill-founded rumor has it that women can't hold their liquor as well as men. Don't bet on it. Always expect a distribution of outcomes including women who can drink you under the table. Heat may have a beneficial effect, but all may not be equally affected. Too many

---

* See Exercise 5 at the end of this chapter.
† Or we may conclude we need more data before we can reach a decision.

other factors can intervene. Any conclusions we draw must be probabilistic in nature. Always, we risk making an error and rejecting a true hypothesis some percentage of the time.*

We say that we make a *type I error* when we accept an alternative hypothesis and yet the null or primary hypothesis is true. We routinely try to reduce the possibility of making a type I error by establishing a significance level (that is, the probability of making a type I error by chance alone) of 5% or even 1%.

We say that we make a *type II error* when we accept the primary hypothesis, yet an alternative is true. *Before* analyzing data, we need to establish a set of values for the test statistic for which we will reject the primary hypothesis; we term this set the *rejection region*. The remaining values of the test statistic for which we will accept the primary hypothesis are collectively termed the *acceptance region*. The boundaries separating these regions are chosen so that the significance level, defined as the probability of making a type I error, will be less than some fixed value.

Our choice also will determine the *power* of the test, the probability of rejecting the hypothesis when a specific alternative is true. The power is 1 minus the probability of making a type II error, and thus depends on the alternative.

After we analyze the data, we will obtain a *p*-value that depends upon the samples. If the *p*-value is less than or equal to the significance level, we will reject the primary hypothesis; otherwise, we will accept it. Note that more experienced researchers will also define a third region, that of indifference, or "I'll need to collect more data and perhaps revise the experimental protocol before I can reach a decision."

Assumptions

The preceding analysis relied on certain assumptions. The *p*-values we obtain by examining all possible rearrangements will be exact, not approximations, *only* if the following is true: *Under the null hypothesis, we can rearrange the labels on the observations without affecting the underlying distribution of possible outcomes.* If this is true, the observations are said to be *exchangeable*.

This requirement deserves a second look. Under the null hypothesis, each possible rearrangement of labels is equally likely. In our previous example, rearrangement

    7.   121 118 12   110 34 22   sum = 251

is as likely as our initial outcome

    1.   121 118 110   34 22 12   sum = 349

---

* A possible exception might arise when we have hundreds of thousands of observations.

under the null hypothesis. But if, to the contrary, heat treatment has a beneficial effect, then our initial outcome is more probable than any other rearrangement, including rearrangement 7.

By *exact*, we mean that when we state that a *p*-value is 5%, it is exactly that. By contrast, *p*-values obtained from the chi-square distribution in the analysis of contingency tables with small numbers of observations can be crude, misleading approximations; a stated 5% might correspond to an actual 20% (see Chapter 8). And when a *t*-test is used to compare two samples of five observations taken from a distribution that is not normal, a stated *p*-value of 5% may correspond to an actual value of 7 or 8%.

When a test is exact, we can guarantee its significance level; that is, we can guarantee that the probability of our making a type I error will be less than some predetermined amount.

The true *p*-values of tests based on methods that make use of tabulated distributions, the so-called parametric methods, such as the chi-square and the *F*-test, depend on the underlying distribution of the observations and are exact only for specific distributions. By contrast, the *p*-values of tests based on permutation methods do not depend on the underlying distribution of the observations and are termed *distribution-free*.

As we saw in Chapter 2, bootstrap confidence intervals are often inaccurate. As we will show in the "Computing the *p*-Value" section, this means that tests based on the bootstrap are not exact. Theoreticians tell us that regardless of the underlying distributions, parametric, permutation, and bootstrap tests of hypotheses are (almost) exact for samples in excess of 100,000 in number. Good news for those who have the money and time to gather extremely large samples.

The ability of permutation tests to detect deviations from the null hypothesis (their *power*) is quite high. This means that we can use smaller samples with permutation tests than some other testing method would require.

General Hypotheses

We can apply permutations whenever we can freely exchange labels under the null hypothesis, as would be the case when testing the hypothesis that both samples are drawn from the same population. Suppose our hypothesis is somewhat different, for example, that we believe a new gasoline additive will increase mileage by at least 20 miles per tank.

Suppose that before using the additive, we recorded 295, 320, 329, and 315 miles per tank of gasoline under various traffic conditions. With the additive, we recorded 330, 310, 345, and 340 miles per tank. To perform a permutation test, we first must recast the primary hypothesis in null form. We do so in this instance by subtracting 20 from each of the postadditive figures. Our two transformed samples are

295 320 329 315

305 290 325 320

Using the transformed mileages, we can test whether the additive increases performance by at least 20 mpg vs. the alternative that the increase in mileage is strictly less than 20 mpg.

Computing the p-Value

Each time you analyze a set of data via the permutation method, you will follow the same five-step procedure:

1. Identify the alternative(s) of interest.
2. Choose a test statistic that best distinguishes between the alternative and the null hypothesis. In the case of the two-sample comparison, the sum of the observations in the sample from the treated population is the obvious choice, as it will be large if the alternative is true and entirely random otherwise.
3. Compute the test statistic for the original labeling of the observations.
4. Rearrange the labels, then compute the test statistic again. Repeat until you obtain the distribution of the test statistic for all possible rearrangements or for a large random sample thereof.
5. Set aside one or two tails of the resulting distribution as your rejection region. In the present example with a one-sided alternative, we would set aside the largest values as our rejection region. If the sum of the treatment observations for the original labeling of the observations is included in this region, then we will reject the null hypothesis.

For small samples, it would seem reasonable (and a valuable educational experience) to examine all possible rearrangements of labels. But for larger samples, even for samples with as few as six or seven values, complete enumeration may be impractical. While there are only 20 different ways we can apply two sets of three labels, there are 924 different ways we can apply two sets of six labels. Though 924 is not a challenge for today's desktop computers, if you are writing your own programs you'll find it simpler and almost as accurate to utilize the *Monte Carlo method*, described next.

Monte Carlo

In a Monte Carlo (named for the Monaco casino), the computer generates random rearrangements of the labels (control and treated; new and old). Suppose that p is the unknown probability that the value of the test statistic for the rearranged values will be as or more extreme than our original value.

Then if we generate n random rearrangements, the number of values that are as or more extreme than our original value will be a binomial random variable $B(n,p)$, with mean np and variance $np(1 - p)$. Our estimate of p based on this random sample of n rearrangements will have an expected value of p and a standard deviation of $[p(1 - p)/n]^{1/2}$.

Suppose p is actually 4% and we examine 400 rearrangements, then 95% of the time we can expect our estimate of p to lie somewhere in the interval between $0.04 - 1.96[.04*.96/400]^{1/2}$ and $.04 + 1.96[.04*.96/400]^{1/2}$, or 2 to 6%. Increasing the number of rearrangements to 6,400 yields a p-value that is accurate to within half of 1%.

R

```
#Two sample comparison of means
N=400 #number of rearrangements to be examined
conventional =c(65, 79, 90, 75, 61, 85, 98, 80, 97, 75)
new = c (90, 98, 73, 79, 84, 81, 98, 90, 83, 88)
n=length (new)
sumorig = sum(new)
cnt= 0 #zero the counter
#Stick both sets of observations in a single vector
A = c(new, conventional)
for (i in 1:N){
  D= sample (A,n)
  if (sum(D) <= sumorig)cnt=cnt+1
}
#increment the count to include the original observations then
compute the pvalue
(cnt+1)/(N+1)
[1] 0.9025
```

SPLUS

Using S+Resample, we may use the R code or do:

```
perm = permutationTest2(new,sum,data2 =
Other,alternative="less")
perm # this prints the p-value, among other things
plot(perm) # plot shows the relationship between the observed
value and the null distribution
```

STATA

Enter the classification as a separate variable.

```
input score method
                score method
   65 0
   79 0
   .  .  ..
   75 0
   90 1
   98 1
   and so forth
```

```
permute score "sum score if method" sum=r(sum), reps(1000)
left nowarn
command: sum score if method
statistic: sum = r(sum)
permute var: score
```

```
Monte Carlo permutation statistics Number of obs = 20
Replications = 1000
```

T	T(obs)	c	n	p=c/n	SE(p)	[95% Conf. Interval]
sum	864	895	1000	0.8950	0.0097	.8743232 .9133152

```
Note: confidence interval is with respect to p=c/n
Note: c = #{T <= T(obs)}
```

Other Two-Sample Comparisons

Two-Sided Test

The preceding section provided program code for a number of one-sided tests. These are the appropriate tests to use when we test whether a new treatment or a new methodology is preferable to an existing one. To fix ideas, suppose we wish to test whether the value to be expected from application of the new treatment is larger than the value to be expected from application of the existing treatment. Our test is based upon the relative values of the means of samples exposed to the new and the existing treatment, respectively. If $\bar{X} < \bar{Y}$, we count the proportion of rearrangements for which $\bar{X}$ is less than or equal to its original value and reject the null hypothesis if this proportion is less than or equal to our predetermined significance level. If $\bar{X} \geq \bar{Y}$, we automatically accept the null hypothesis.

On the other hand, if we wish to test the null hypothesis the population mean of X is identical with the population mean of Y against the two-sided alternative that the population mean is unequal, we would proceed as follows:

- If $\bar{X} < \bar{Y}$, we count the proportion of rearrangements for which $\bar{X}$ is less than or equal to its original value, and reject the null hypothesis if twice this proportion is less than or equal to our predetermined significance level.
- If $\bar{X} \geq \bar{Y}$, we count the proportion of rearrangements for which $\bar{X}$ is greater than or equal to its original value, and reject the null hypothesis if twice this proportion is less than or equal to our predetermined significance level.

The resultant p-value correctly reflects our intent to reject if the value of our test statistic proves to be either extremely large or extremely small values with respect to the distribution of all possible values that might be obtained by random relabeling.

Rank Tests

Back in the day when the basement of Campbell Hall was occupied by a computer that would spend the entire night cranking away on a single two-sample permutation test, the use of ranks rather than the original observations offered the advantage that the cutoff value for one pair of samples of sizes n and m held for all similarly sized pairs. Thus were born the rank tests: the Wilcoxon and the Kruskal-Wallace.

With today's high-speed microcomputers, one can use the original observations rather than their ranks, for the p-value for a two-sample permutation appears on the screen at almost the same instant we enter the data. If we do use ranks rather than the original data means, we'll need to take larger samples to make up for the loss in sensitivity (power). In the most favorable case with very large samples, a rank test requires 100 observations for every 97 using the original observations.

Rank tests should be used *only* in the following three instances:

1. One or more extreme values are suspect (are they typographical errors?).
2. The methods used to make the measurement were not the same for each observation.
3. The measurements were made on different scales.

Matched Pairs

We can increase the power of our tests by eliminating or reducing unwanted sources of variation. One of the best ways to eliminate a source of variation and subsequent errors is through the use of *matched pairs*. Each subject in one group is matched as closely as possible by a subject in the treatment group. If a 45-year-old black male hypertensive is given a blood-pressure-lowering pill, then we give a second similarly built 45-year-old black male hypertensive a placebo.

Suppose that each of three subjects in a clinical trial served as his or her own control. We would permute the labels "control" and "treatment" separately and independently for each of the three.

Original Results

Subject	1	2	3
Control	17	19	14
Treated	21	21	17

Rearrangement One

Subject	1	2	3
Control	21	19	14
Treated	17	21	17

Rearrangement Two

Subject	1	2	3
Control	17	21	14
Treated	21	19	17

Rearrangement Three

Subject	1	2	3
Control	21	21	14
Treated	17	19	17

With three subjects, eight rearrangements are possible. With n, we would have 2^n possible rearrangements. Using as our test statistic the sum of the control observations, we would reject only if this sum, when calculated for our observations with their original labels, fell in a tail of the permutation distribution.

R Code

```
New = c(48722, 28965, 36581, 40543, 55423, 38555, 31778, 45643)
Standard=c(46555, 28293, 37453, 38324, 54989, 35687, 32000,
43289)
Diff=New-Standard
```

```
N=400 #number of rearrangements to be examined
sumorig = sum(Diff)
n=length(Diff)
stat = numeric (n)
cnt= 0 #zero the counter
for (i in 1:N){
for (j in 1:n) stat[j]=ifelse(runif(1) < 0.5, Diff[j],-Diff[j])
if (sum(stat) >= sumorig)cnt=cnt+1
}
#increment the count to include the original observations then
compute the pvalue
(cnt+1)/(N+1)
[1] 0.032
```

Stata

```
drop _all
input new stand
48722 46555
28965 28293
36581 37453
40543 38324
55423 54989
38555 35687
31778 32000
45643 43289
end

set seed 1234
local reps 400
tempvar which
quietly gen `which' = 0
sum new, meanonly
scalar sumorig = r(sum)
local cnt 0
forval i = 1/`reps' {
quietly replace `which' = cond(uniform() < 0.5, new, stand)
sum `which', meanonly
if r(sum) <= scalar(sumorig) {
local ++cnt
}
}
di `cnt'/`reps'
exit
```

Test for Nonequivalence

Suppose we wish to test the hypothesis that two formulations are equivalent

$$\mu_G - \delta_L \; <= \; \mu_N <= \mu_G + \delta_U$$

against the alternative hypothesis that they are not equivalent.

We would draw samples from each process and examine $S = \text{Mean}(N) - \text{Mean}(G)$.

$$\mu_N <= \mu_G - \delta_L \text{ or } \mu_N >= \mu_G + \delta_U$$

If $\delta_L \le S \le \delta_U$, we accept the hypothesis.

If $S \le \delta_L$, we add δ_L to each value in the sample of New. We compute the permutation distribution of S, and reject the hypothesis if the original value of S is less than or equal to the $\alpha/2$th percentile of this distribution.

If $S \ge \delta_U$, we subtract δ_U from each value in the sample of New. We compute the permutation distribution of S, and reject the hypothesis if the original value of S is greater than or equal to the $(100 - \alpha/2)$th percentile of this distribution.

On the down side, our acceptance of the hypothesis in this instance may just be the result of an inadequate sample size. The bootstrap described in the "Balanced Bootstrap" section in Chapter 2 may be preferable.

Underlying Assumptions

At this point we are in a position to distinguish among the various hypothesis testing procedures—parametric, permutation, and bootstrap—on the basis of their underlying assumptions.

To use any of the tests, the observations must be independent.

To use either a permutation or a parametric test, the observations must be exchangeable; that is, their joint distribution under the null hypothesis remains unchanged when the labels are rearranged.

To use a parametric test, the observations must all come from a distribution of predetermined form. As far as continuous measurements are concerned (height, weight, blood pressure), the two-sample comparison is also the exception that proves the rule. As noted earlier, an estimator that is the sum of a large number of independent values, each of which contributes only a small amount to the total sum, will have the normal distribution. If we have sufficient observations, then their mean will be normally distributed regardless of the distribution of the observations themselves. Thus, Student's *t*-test is recommended for comparing the means of two distributions for samples of 10 or more.

Comparing Variances

Precision is essential in a manufacturing process. Items that are too far out of tolerance must be discarded. An entire production line is brought to a halt if too many items exceed (or fall below) designated specifications. With some testing equipment, such as that used in hospitals, precision can be more important than accuracy. Accuracy can always be achieved through the use of standards with known values, while a lack of precision may render an entire sequence of tests invalid.

Though many methods are available with which to test the hypothesis that two samples come from populations with the same inherent variability, few can be relied on. Sukhatme (1958) lists four alternative approaches and adds a fifth of his own. Miller (1968) lists 10 alternatives and compares 4 of these with a new test of his own. Conover et al. (1981) list and compare 56 tests. Balakrishnan and Ma (1990) list and compare nine tests, with one of their own. Many promise a type I error rate of 5%, but in reality make errors as frequently as 8 to 20% of the time. Other methods for comparing variances have severe restrictions.

For example, a permutation test based on the ratio of the sample variances is appropriate only if the means of the two populations are the same or if we know their values. If the means are not the same, then the labels can not be exchanged unless we know the values of the means; in such a (rare) case, we can make a preliminary transformation that makes the labels exchangeable.

We can derive a permutation test for comparing variances that is free of these restrictions if, instead of working with the original observations, we replace them with the differences between successive order statistics and then permute the labels. The test statistic proposed by Aly (1990) is

$$\delta = \sum_{i=1}^{m-1} i(m-i)(X_{(i+1)} - X_{(i)})$$

where $X_{(1)} \le X_{(2)} \le \ldots \le X_{(m)}$ are the order statistics of the first sample. That is, $X_{(1)}$ is the smallest of the observations in the first sample (the minimum), $X_{(2)}$ is the second smallest, and so forth, up to $X_{(m)}$, the maximum.

To illustrate the application of Aly's statistic, suppose the first sample consists of the measurements 121, 123, 126, 128.5, and 129, and the second sample the measurements 153, 154, 155, 156, and 158. $X_{(1)} = 121$, $X_{(2)} = 123$, and so forth.

Set $z_{1i} = X_{(i+1)} - X_{(i)}$ for $i = 1, \ldots, 4$. In this instance, $z_{11} = 123 - 121 = 2$, $z_{12} = 3$, $z_{13} = 2.5$, $z_{14} = 0.5$.

The original value of Aly's test statistic is $8 + 18 + 15 + 2 = 43$. To compute the test statistic for other arrangements, we also need to know the differences $z_{2i} = Y_{(i+1)} - Y_{(i)}$ for the second sample; $z_{21} = 1$, $z_{22} = 1$, $z_{23} = 1$, $z_{24} = 2$.

Only certain exchanges are possible. Rearrangements are formed by first choosing either z_{11} or z_{21}, next either z_{12} or z_{22}, and so forth until we have a set of four differences.

One possible rearrangement is {2, 1, 1, 2}, which yields a value of δ = 24. There are $2^4 = 16$ rearrangements in all, of which only one {2, 3, 2.5, 2} yields a more extreme value of the test statistic than our original observations. With 2 out of 16 rearrangements yielding values of the statistic as or more extreme than the original, we should accept the null hypothesis. Better still, given the limited number of possible rearrangements, we should gather more data before we make a decision.*

R Code for Aly's Test Statistic

```
diff=function(samp){
   s=sort(samp)
   l= length(samp)
   d=1:(l-1)
   for(k in 2:l){
      d[k-1]=s[k]-s[k-1]
   }
   return(d)
}

aly=function(samp){
   stat=0
   l=length(samp)
   for (k in 1:l) stat=stat+k*(l+1-k)*samp[k]
   return(stat)
}

vartest=function(samp1,samp2, NMonte){
   d1=diff(samp1)
   d2=diff(samp2)
   l=length(d1)
   stat0=aly(d1)
   pd=d1
   cnt=0
   for(j in 1:NMonte){
      r=rbinom(l,1,.5)
      for (k in 1:l)pd[k]=ifelse(r[k],d1[k],d2[k])
      if (aly(pd)>=stat0)cnt=cnt+1
   }
   return(cnt/NMonte) #one-sided p-value
}
x1 = c(129, 123, 126, 128.5, 121)
y1 = c(153, 154, 155, 156, 158)
vartest(x1,y1,1600)
[1] 0.20125
```

---

* How much data? See Chapter X.

A word of caution: If we are performing a two-sided test and the *p*-value, defined as the ratio of the number of more extreme values (cnt) to the number of simulations (Nmonte), obtained from the preceding program is less than 0.5, we will need to double it to be consistent with our intention to reject if the test statistic proves to be either extremely large or extremely small. If the *p*-value is greater than 0.5, this suggests that the second sample is less dispersed than the first. Consequently, we will need to rerun vartest() after modifying the appropriate line of its listing to read

```
if (aly(pd)<=stat0)cnt=cnt+1.
```

We also would need to double the resulting *p*-value.

Unequal Sample Sizes

If our second sample is larger than the first, we have to resample in two stages. Suppose m observations are in the first sample and n in the second, where $m < n$. Select a random subset of m values $\{Y_{*i}, i = 1, \ldots, m\}$ without replacement from the n observations in the second sample. Compute the order statistics $Y_{*(1)} \leq Y_{*(2)} \leq \ldots$, their differences $\{z_{*2i}\}$, and the value of Aly's measure of dispersion for the 2^m possible rearrangements of the combined sample $\{\{z_{1i}\},\{z_{*2i}\}\}$. Repeat this procedure for all possible subsets of the second sample, combine all the permutation distributions into one, and compare Aly's measure for the original observations with the combined distribution.

Should the total number of calculations appear prohibitive were all subsets used, then use only a representative random sample of them.

Preferred Method

Alas, Aly's test for unequal sample sizes lacks power, and it is seldom that we finish an experiment in which the sample sizes have remained equal. In the first edition of *Permutation Tests*, Good (1994) proposed a permutation test based on the sum of the absolute values of the deviations. First, we compute the median for each sample; next, we replace each of the remaining observations by the square of their deviations about their sample median; last, in contrast to the test proposed by Brown and Forsythe (1974), we discard the redundant linearly dependent value from each sample.

Suppose the first sample contains the observations $x_{11}, \ldots, x_{1n}$, whose median is M_1; we begin by forming the deviates $x'_{1j} = |x_{1j} - M_1|$ for $j = 1, \ldots, n_1$. Similarly, we form the set of deviates $\{x'_{2j}\}$ using the observations in the second sample and their median.

If there are an odd number of observations in the sample, then one of these deviates must be zero. We can't get any information out of a zero, so we throw it away. In the event of ties, should there be more than one zero, we still throw only one away. If there is an even number of observations in the sample, then two of these deviates (the two smallest ones) must be equal. We can't get any information out of the second one that we didn't already get from the first, so we throw it away.

Our new test statistic S_G is the sum of the remaining $n_1 - 1$ deviations in the first sample; that is, $S_G = \sum_{j=1}^{n_1-1} x'_{1j}$.

We obtain the permutation distribution for S_G and the cutoff point for the test by considering all possible rearrangements of the remaining deviations between the first and second samples.

To illustrate the application of this method, suppose the first sample consists of the measurements 121, 123, 126, 128.5, and 129.1, and the second sample the measurements 153, 154, 155, 156, and 158. Thus, after eliminating the zero value, $x'_{11} = 5$, $x'_{12} = 3$, $x'_{13} = 2.5$, $x'_{14} = 3.1$, and $S_G = 13.6$. For the second sample $x'_{21} = 2$, $x'_{22} = 1$, $x'_{23} = 1$, and $x'_{24} = 3$.

All of the $\binom{8}{4}$ possible arrangements yield only three values of the test statistic as or more extreme than our original value; $3/70 = 0.043$, and we conclude that the difference between the dispersions of the two manufacturing processes is statistically significant at the 5% level.

As there is still a weak dependency among the remaining deviates within each sample, they are only asymptotically exchangeable. Tests based on S_G are alternately conservative and liberal according to Baker (1995), in part because of the discrete nature of the permutation distribution, unless:

1. The ratio of the sample sizes n, m is close to 1.
2. The only other difference between the two populations from which the samples are drawn is that they might have different means, that is, $F_2[x] = F_1[(x - \delta)/\sigma]$.

We were unable to confirm her results in our own simulations (the R code for which may be obtained from us). We found tests based on S_G to be uniformly conservative for samples of sizes 4 and above. Our simulations employed either normally distributed data or mixed normal data. To avoid randomizing on the boundary in our simulations, we set the alpha level in each instance to correspond to one of the discrete levels available for the permutation distribution. Chernick and Liu (2002) describe the necessity of such a procedure.

In Chapter 6, we show how this latter approach may be extended to k-sample comparisons as well as to mixtures of responders and nonresponders.

R Code

```
#eliminate smallest(and redundant)residual
resid=function(data){
  mn=median(data)
  res=abs(data-mn)
```

```
res=sort(res)
return (res[-1])
}

ptest=function(res, sizeS, MC){
  sumorig=sum(res[1:sizeS])
  cnt=0
  for (i in 1:MC){
    D=sample(res,sizeS)
    if (sum(D)<= sumorig)cnt<-cnt+1
    }
  return(cnt/MC)
  }
```

Testing in the Presence of Nonresponders

In testing for a response to drug treatment, it is common to encounter a response threshold peculiar to each individual, such that some individuals respond to drug treatment and some do not. If the treatment is effective, one expects both the mean and the variance of the treated population to be larger than those of the control population. This suggests that a test for simultaneous changes in expectation and variance would be more powerful than one that tests for changes in expectation alone.

We wish to test the hypothesis H: $F_2[x] = F_1[x]$ against the alternative K:

$$F_2[x] = pF_1[x] + (1-p) F_1[(x-d)/s]; 0 < p < 1; d > 0 ; s \geq 1.$$

Good (1979) proposed the test statistic

$$v = u(\bar{X}_T - \bar{X}_C)^2 + (1-u)S_T^2$$

where the first term is proportional to the difference in means of the two samples and the second to the variance of the treatment sample. Rearranging the labels between the two sets of observations generates its permutation distribution. But alas, its power is only marginally better than the *t*-test.

We suggest instead the test statistic $T = u(\bar{X}_\pi - \bar{X}_O) + (1-u)(S_\pi - S_O)$, where S is the sum of the deviations about the median as defined in the previous section, and the subscripts O and π refer to the original data and the data after rearranging sample labels.

Care must be taken in generating the rearrangements, as the first part of our test statistic is based on one more value than the second. To accomplish the desired result, we first select $n-1$ observations at random from the reduced data set to use in forming S, and then one more observation (of those not already selected) to calculate the mean.

Ideally, the parameter u would be chosen equal to p, but typically p is not known. It may be estimated using the method of Qin and Liang (2010). In our simulations, we used a value of $u = 0.67$ and selected data from an $N(1, 1)$ population for controls and from a mixture of 50% $N(1, 1)$ and 50% $N(2, 2)$ for the treated group. At a significance level of 10%, and using two samples of size 5, respectively, the t-test yielded power of 20%, a permutation test using Student's t as its test statistic had a power of 21%, and a permutation test that made use of our new statistic had a power of 33%.

Summary

In this chapter you were provided with a method for obtaining exact distribution-free tests based on any statistic, providing that the treatment labels could be exchanged among the observations without affecting their joint probability if the null hypothesis were true. Thus, by making use of permutation tests, you are free to choose the statistic that best discriminates between the null hypothesis and alternatives of practical interest.

To illustrate the approach, you were provided with algorithms for comparing the means and variances of two samples and shown you could make use of the original observations or their ranks. You saw that the R code of the "Computing the p-Value" section could be generalized as follows:

```
N=1600 #number of rearrangements to be examined
statorig = statistic_of_your_choice(sample)
cnt= 0 #zero the counter
for (i in 1:N){
   D= sample (A,)
   if (statistic_of_your_choice (D) <= statorig) cnt=cnt+1
}
cnt/N #pvalue
```

To Learn More

The permutation tests were introduced by Pitman (1937, 1938) and Fisher (1935). Texts dealing with their application include Bradley (1968), Edgington and Onghena (2007), Noreen (1989), and Manly (2006). Early articles include Pearson (1937) and Wald and Wolfowitz (1944). Robust permutation tests are considered by Maritz (1996) and Huang et al. (2009).

Optimal tests for the analysis of group randomized trials are derived by Braun and Feng (2001).

Advanced applications of permutation methods will be found in Chapters 5 to 7.

For a primer on the application of permutation methods to PET images with many worked through examples, see Nichols and Holmes (2001).

Exercises

1. Suppose my neighbor had been given two more observations: 90 and 95. What would be the results of a permutation analysis applied to the new, enlarged data set consisting of the following values?

   Treated: 121 118 110 90

   Untreated: 95 34 22 12

   a. Solve this problem by determining both the total number of possible rearrangements and the number of rearrangements that are as or more extreme than the one observed.

   b. Solve this problem by running a Monte Carlo analysis.

   c. What would the p-value be if you also were concerned with the possibility that heat treatment might decrease tensile strength.

2. Do men and women have the same views on health care reform? Use the data from http://www.statcrunch.com/5.0/index.php?dataid= 402742. Be sure to block by the respondent's political party.

3. How would you go about testing the hypothesis that the fuel additive described in the "General Hypotheses" section increases mileage by 10%?

4. Using the data from http://www.statcrunch.com/5.0/index.php? dataid=376457:

   a. Determine whether the treatment had an effect.

   b. Estimate the magnitude of the effect. (Hint: See Chapter 2.)

5. One hundred forty-four hours after mice were inoculated with Herpes virus type II, the following virus titers were observed in their vaginas (see Good, 1979 for complete details):

   Saline controls: 10,000, 3,000, 2,600, 2,400, 1,500

   Treated with antibiotic: 9,000, 1,700, 1,100, 360, 1

   a. Does treatment have an effect?

   b. Most authorities would suggest using a logarithmic transformation before analyzing these data because of the exponential nature of viral growth. Repeat your analysis after taking the logarithm of each observation. Is there any difference? Compare your results and interpretations with those of Good (1979).

6. You can't always test a hypothesis. A marketing manager would like to show that an intense media campaign just before Christmas resulted in increased sales. Should he compare this year's sales with last year's? What would the null hypothesis be? And what would be some of the alternatives?

7. An economist developed the following model for personal consumption expenditures C as a function of per capita disposable income D:

$$C[i] = 445 + 0.85D[i] + e[i]$$

where the model errors $e[i]$ are independent and identically symmetrically distributed with a mean of zero. Test the hypothesis that the model is correct, that is, the expected value of $C = 445 + 0.85D$, against the alternative that the expected value of $C > 445 + 0.85D$ with the aid of the following data:

$$D = 6{,}036,\ 6{,}113,\ 6{,}271,\ 6{,}378,\ 6{,}727,\ 7{,}027,\ 7{,}280$$

$$C = 5{,}561,\ 5{,}579,\ 5{,}729,\ 5{,}855,\ 6{,}099,\ 6{,}362,\ 6{,}607$$

8. This final exercise illustrates the value of using the most powerful tests. Derado et al. (2004) performed a series of complex time-consuming measurements on 12 difficult to obtain and house monkeys, when 6 monkeys would have yielded the same result had they used a permutation test to analyze the results instead of bootstrap methods. To see why, analyze the results below using both permutation methods and the bootstrap. Repeat the tests using only six of the monkeys chosen at random.

| Before | 0.212 | 0.089 | 0.133 | 0.143 | 0.154 | 0.165 | 0.084 | 0.135 | 0.143 | 0.117 | 0.115 | 0.287 |
| After | 0.245 | 0.130 | 0.202 | 0.257 | 0.178 | 0.215 | 0.144 | 0.160 | 0.154 | 0.102 | 0.114 | 0.320 |

5

Multiple Variables

The value of an analysis based on simultaneous observations on several variables, such as height, weight, blood pressure, and cholesterol level, is that it can be used to detect subtle changes that might not be detectable except with very large, prohibitively expensive samples, were we to consider only one variable at a time.

The resampling methods are of particular value in a multivariate setting, as the parametric requirement of multivariate normality is almost never satisfied. You will learn two approaches to multivariate analysis in this chapter:

- Choosing a single-valued test statistic that can stand in place of the multivalued vector of observations
- Combining the p-values associated with the various univariate tests into a single p-value

Single-Valued Test Statistic

Hotelling's T^2

Hotelling's T^2 is a straightforward generalization of Student's t to multiple variables per observation. Some notation is necessary. We use bold type to denote a vector $\mathbf{X}$ whose components $(X_1, X_2, \ldots, X_J)$ have expected values $\mu = (\mu_1, \mu_2, \ldots, \mu_J)$. X_1 might be a student's GPA, X_2 the same student's SAT score, and so forth. If we have collected vectors of observations for n students, then let $\bar{\mathbf{X}}$ denote the corresponding $1 \times I$ vector of sample means, and $\mathbf{V}$ the matrix whose ijth component is the covariance of X_i and X_j.

In the one-sample case, Hotelling's T^2 is defined as $(\bar{\mathbf{X}} - \mu)V^{-1}(\bar{\mathbf{X}} - \mu)^T$.

In the two-sample case, Hotelling's T^2 is defined as $(\bar{\mathbf{X}}_1 - \bar{\mathbf{X}}_2)V^{-1}(\bar{\mathbf{X}}_1 - \bar{\mathbf{X}}_2)^T$, where the ijth component of $\mathbf{V}$ is estimated from the combined sample by the formula

$$V_{ij} = \frac{1}{n_1 + n_2 - 2} \sum_{g=1}^{2} \sum_{k=1}^{n_g} (x_{gik} - \bar{x}_{gi.})(x_{gjk} - \bar{x}_{gj.})$$

This statistic weighs the contribution of individual variables and pairs of variables in inverse proportion to their covariances. This has the effect of rescaling each variable so that the most weight is given to those variables that can be measured with the greatest precision, as well as those that can provide information not provided by the others.

As the means of measurements often have a near-normal distribution, Student's *t* is recommended for most two-sample univariate comparisons. In contrast, the significance level of the multivariate Hotelling's T^2 is highly sensitive to departures from normality, and a resampling approach is recommended (see Davis, 1982).

To perform a permutation test, we treat each vector of observations on an individual subject as a single indivisible entity. When we exchange treatment labels, we relabel on a subject-by-subject basis so that all observations on a single subject receive the same new treatment label. In other words, we exchange labels among the rows of the matrix of observation, so that all the observations on a single subject receive the same label.

We proceed in three steps:

- We compute Hotelling's T^2 for the original observations.
- We compute Hotelling's T^2 for all relabelings of the subjects.
- We determine the percentage of relabelings that lead to values of the test statistic that are as or more extreme than the original value.

Note that we are forced to compute the covariance matrix **V** and, more time-consuming, its inverse for each new rearrangement. To reduce the number of computations, Wald and Wolfowitz (1944) suggest using a slightly different statistic *W* that is large when T^2 is large and vice versa.

$$W = (\bar{\mathbf{X}}_1 - \bar{\mathbf{X}}_2)C^{-1}(\bar{\mathbf{X}}_1 - \bar{\mathbf{X}}_2)^T$$

where the components of *C* are

$$c_{ij} = \sum_{g=1}^{2} \sum_{k=1}^{n_g} (x_{gik} - U_i)(x_{gjk} - U_j)$$

with

$$U_i = \frac{1}{n_1 + n_2} \sum_{g=1}^{2} \sum_{k=1}^{n_g} x_{gik}$$

Complicated? As always, we let the computer do the calculations for us. In this example, our objective is to see whether there are significant differences between foreign and domestic automobiles in terms of price, miles per gallon, headroom, and turning radius. Suppose we load the data set auto.dta into Stata.

Entering the Stata command

```
. permute foreign "hotelling price mpg turn headroom, by
(foreign)" r(T2), reps(400)
```

yields the following output:

```
command: hotelling price mpg turn headroom, by(foreign)
statistic: _pm_1 = r(T2)
permute var: foreign
Monte Carlo permutation statistics Number of obs = 74
Replications = 400
T T(obs) c n p=c/n SE(p) [95% Conf. Interval]
_pm_1 62.52314 0 400 0.0000 0.0000 0.0091798
```

The 95% confidence interval is with respect to $p = c/n$, where c is the number of rearrangements for which $T \geq T(obs)$. In this example of 400 rearrangements, none yielded a value of Hotelling's T^2 as large as the value associated with the original observations. The probability of this happening by chance alone is less than 1%.

Application to Repeated Measures

Although successive observations *must* be independent for most statistical methods to be applicable, the variables that make up an observation may be and often are interdependent. For example, in a time-course experiment, when we take a series of measurements on the same subject, these measurements are *not* independent. Suppose you then use one of the regression techniques described in Chapter 10 to derive a best-fitting curve for each patient. Following Zerbe and Walker (1977), you would replace each set of measurements taken over time with the corresponding set of regression coefficients. The problem of a two-treatment comparison reduces to that of a multivariate comparison between the regression coefficients of the two treatment groups, and you may use the methods described in the preceding section to obtain the permutation distribution of Hotelling's T^2.

Three distinct questions about the time-course profiles may be addressed:

1. Are the profiles the same for the various treatments?
2. Are the profiles parallel?
3. Are the response profiles at the same level?

A yes answer to question 1 implies yes answers to questions 2 and 3, but we may get a yes answer to 2 even when the answer to 3 is no.

One simple test of parallelism entails computing the successive differences in value from time point to time point and then applying the methods of the preceding section to these differences. Of course, this approach is applicable only if the observations on both treatments were made at identical times.

To circumvent this limitation and to obtain a test of the narrower hypothesis of equivalent response profiles, Koziol et al. (1981) use an approach based on ranks: Suppose there are n_g subjects in the gth treatment group and n_{gt} observations were made on these subjects at time t. Let R_{gjt} be the rank of the observation on the jth subject among these n_{gt} values.

If luck is with us so that all subjects remain to the end of the experiment, then $n_{gt} = n_g$ for all t and each treatment group, and we may adopt the test statistic first proposed by Puri and Sen (1966):

$$L = \sum_g n_g \bar{\mathbf{R}}_g V^{-1} \mathbf{R}_g^{\ T},$$

where $\bar{\mathbf{R}}_g$ is a $1 \times T$ vector whose tth component is $\bar{R}_{gt}$ and V is a $T \times T$ covariance matrix whose stth component is $\Sigma_g \Sigma_k R_{gkt} R_{gks}$.

The Generalized Quadratic Form

Originally proposed for use in epidemiology, Mantel's $U = \Sigma\Sigma a_{ij} b_{ij}$ has a broad range of applications. In Mantel (1967) a_{ij} is a measure of the difference in time between items i and j, while b_{ij} is a measure of the spatial distance. The actual values of the coefficients will depend upon the specific application. As an example, suppose t_i is the day on which the ith individual in a study came down with cholera and (x_i, y_i) are the coordinates of her position in space (for example, five blocks west and two blocks north of city center). For all i, j one could set $a_{ij} = 1/(t_i - t_j)$ and $b_{ij} = 1/\sqrt{(x_i - x_j)^2 + (y_i - y_j)^2}$.

A large value for U would support the view that cholera spreads by contagion from one household to the next. To obtain the permutation distribution of U and assess whether or not its observed value is statistically significant, we fix the times i but permute the distances j, as in $\pi[U] = \Sigma\Sigma a_{ij} b_{i\pi[j]}$.

The definition of Mantel's U can be seen to include several of the standard measures of correlation, including those usually attributed to Pearson, Pitman, Kendall, and Spearman (Hubert, 1985). See Exercise 3.

Application to Epidemiology

Many factors in our environment—asbestos or radon in the walls of our houses, or toxic chemicals in our air and groundwater—will affect our offspring. Siemiatycki and McDonald (1972) studied the incidence of congenital neural tube defects; their results are shown in Table 5.1. Note the gradient along the diagonal of this table, from which one might infer that

TABLE 5.1

Incidents of Pairs of Ancephalic Infants

km Apart	Months Apart		
	<1	1< <2	2< <4
<1	39	101	235
<5	53	156	364
<25	211	652	1,516

births of ancephalic infants occur in clusters. We could test this hypothesis statistically using the methods of the "Ordered Statistical Tables" section for ordered categories, but a better approach, since the exact time and location of each event are known, is to use Mantel's U. The result will depend on the measures of distance and time we employ. Mantel (1967) reports striking differences between an analysis of epidemiologic data in which the coefficients are proportional to the differences in position and a second approach (which he recommends) to the same data in which the coefficients are proportional to the reciprocals of these differences.

Applying this latter approach, a pair of infants born 5 km and 3 months apart contribute $1/3*1/5 = 1/15$ to the statistic. Summing up the contributions from all pairs, then repeating the summing process for a series of random rearrangements, Siemiatycki and McDonald (1972) conclude the clustering of ancephalic infants is not statistically significant. You can confirm their result by entering the data into StatXact, using "TableData, Settings" to specify the row weights (2, 0.4, 0.08) and the column weights (2, 0.67, 0.33), then selecting "Statistics," "Doubly Ordered R x C Table," " Linear-by-linear," and "Exact" from the menus to obtain a p-value of 0.18.

Further Generalization

The coefficients of Mantel's U need not correspond to space and time. Hubert and Schultz (1976) studied k distinct variables in each of a large number of subjects in order to test a specific sociological model. The $\{a_{ij}\}$ in Mantel's U were elements of the $k \times k$ sample correlation matrix, while the $\{b_{ij}\}$ were elements of an idealized or theoretical correlation matrix derived from the model. A large value of U supported the model; a small value would have ruled against it.

The MRPP Statistic

Mielke's (1986) multiresponse permutation procedure has been applied to applications as diverse as the weather and the spatial distribution of archaeological artifacts. The MRPP uses the permutation distribution of between-object distances to determine whether a classification structure has a nonrandom distribution in space or time. The method is flexible enough that it is readily adapted to obtain a permutation version of Hotelling's T^2 comparison of multivariate medians.

An example of MRPP's application arises in the assignment of antiquities to specific classes based on their spatial locations in an archaeological dig. Presumably, the kitchen tools of primitive man—woks and Cuisinarts— should be found together, just as a future archaeologist can expect to find a TV, DVD player, and home theater side by side in a 21st-century living room.

Suppose we have a collection of N artifacts within a site, and the artifacts can be divided into G distinct groups (dishes, electronic devices, knick-knacks, and not yet classified), with n_g artifacts tentatively assigned to the gth class. Following Berry et al. (1980, 1983), let D_g denote the average distance between artifacts within the gth class. The test statistic is the weighted within-class average of these distances, $D = \Sigma_g n_g D_g / N$. The permutation distribution associated with D is taken over all allocations of the N artifacts to the G classes, with n_g assigned to each class.

Analyzing Migration Data

The distance and elevation change (in meters) for male and female blue grouse migrating from where they were marked on their breeding range to their winter range are given in Table 5.2, taken from Cade and Hoffman (1993). Generally the males seem to migrate farther and higher than the females. Suppose our primary hypothesis is that the migration patterns of males and females are the same, and the alternative of interest is that males do fly farther and higher than females. The parametric version of Hotelling's T^2 based on gender differences in both distance and elevation yields a p-value of 0.033. The permutation test yields a p-value of 0.003.

To obtain a p-value based on the permutation distribution of Hotelling's T^2, using the Blossom software (see Chapter 3) you would issue the following commands:

```
USE BGROUSE.DAT
>MRPP DIST ELEV * SEX/HOT V = 2 C = 2 EXACT
```

TABLE 5.2

Blue Grouse Migration Data

Distance (km)	8	64	78	105	106	118	125	141	174	294	1
Elevation	0	503	488	457	610	183	549	549	671	427	0
Sex	F	M	M	M	M	M	M	M	M	M	F
Distance	1	4	7	13	52	44	67	88	151	280	
Elevation	76	198	91	213	320	−61	305	518	360	760	
Sex	F	F	F	F	F	F	F	F	F	F	

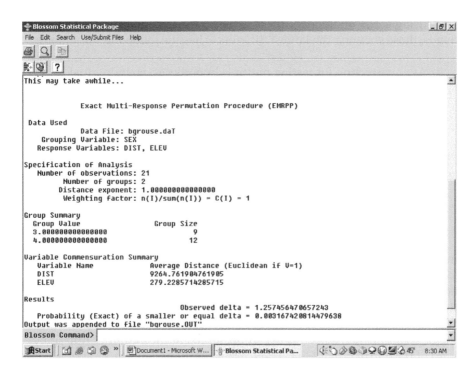

FIGURE 5.1
Result of applying MRPP to the blue grouse data.

This latter command bases its calculations on the square of the Euclidean distances. Mielke and Berry (1999) showed that one can obtain a more powerful test by using the absolute value of the Euclidean distance via the Blossom command:

```
>MRPP DIST ELEV * SEX/EXACT
```

which yields the output shown in Figure 5.1.

Gene Set Enrichment Analysis

The use of permutation methods frees us to make use of the statistic most appropriate to our application, even if tabulated values of the statistic are not available. As one example, we describe in this section a multivariate procedure that is specific to the analysis of microarrays.

Typically, mRNA expression profiles are generated for thousands of genes from a collection of samples belonging to one of two outward expressions of the genes, for example, tumors that are sensitive vs. resistant to a drug. The genes can be ordered in a ranked list according to their differential expression between the classes.

Focusing solely on the handful of genes at the top and bottom of this list, that is, those showing the largest difference, has the following limitations:

1. After correcting for testing multiple hypotheses, no individual gene may meet the threshold for statistical significance.
2. Or, one may be left with a long list of statistically significant genes without any unifying biological theme.
3. Or, single-gene analysis may miss important effects on pathways. Cellular processes often affect sets of genes acting in concert. An increase of 20% in all genes encoding members of a metabolic pathway may dramatically alter the flux through the pathway and may be more important than a 20-fold increase in a single gene.

Given an a priori defined set of genes S (that is, genes encoding products in a metabolic pathway, located in the same cytogenetic band, or sharing the same gene ontology category), the goal of gene set enrichment analysis is to determine whether the members of S are randomly distributed throughout the ranked list or primarily found at its top or bottom.

Subramanian et al. (2005) calculate an enrichment score (ES) that reflects the degree to which a set S is overrepresented at the extremes (top or bottom) of the entire ranked list. The score is calculated by walking down the list, increasing a running sum statistic when a gene in S is encountered, and decreasing it when we encounter genes not in S. The magnitude of the increment depends on the correlation of the gene with the phenotype. The enrichment score is the maximum deviation from zero encountered in the random walk; it corresponds to a weighted Kolmogorov-Smirnov-like statistic.

They determine the permutation distribution of ES by permuting the phenotype labels and recomputing the ES of the gene set for the permuted data. Permuting the class labels preserves gene-gene correlations, providing a more biologically reasonable assessment of significance than would be obtained by permuting genes.

Combining Univariate Tests

We also can make a comparison of data derived from interdependent observations by combining the univariate tests. Higgins and Noble (1993) analyze an experiment whose goal was to compare two methods of treating beef carcasses in terms of their effect on pH measurements of the carcasses taken over time. Treatment level B is suspected to induce a faster decay of pH values. Formally, they test a hypothesis of no difference between the treatments

against the alternative that $X_{B[t]}$ is stochastically smaller than $X_{A[t]}$ for some time t and no greater at other times.
   Observed data are:

$t = 0\ 1\ 2\ 3\ 4\ 5$

A1 6.81 6.16 5.92 5.86 5.80 5.39

A2 6.68 6.30 6.12 5.71 6.09 5.28

A3 6.34 6.22 5.90 5.38 5.20 5.46

A4 6.68 6.24 5.83 5.49 5.37 5.43

A5 6.79 6.28 6.23 5.85 5.56 5.38

A6 6.85 5.51 5.95 6.06 6.31 5.39

B1 6.64 5.91 5.59 5.41 5.24 5.23

B2 6.57 5.89 5.32 5.41 5.32 5.30

B3 6.84 6.01 5.34 5.31 5.38 5.45

B4 6.71 5.60 5.29 5.37 5.26 5.41

B5 6.58 5.63 5.38 5.44 5.17 5.62

B6 6.68 6.04 5.62 5.31 5.41 5.44

   Although normality of these observations may be assumed, the variances and covariances surely vary with time so that the two-way ANOVA model is not appropriate. Instead, one may proceed as follows: First, standardize the observations, subtracting the baseline value at $t = 0$ from each one. At each time point, the resulting differences are exchangeable. Treating each time point separately, the resulting p-values are as follows:

$T = 1$	$T = 2$	$T = 3$	$T = 4$	$T = 5$
0.01056	0.000127	0.000309	0.000395	0.06803

   Using Fisher's nonparametric combination rule, $F = -2\log[\Pi_i p_i] = -2\Sigma_i \log[p_i]$, the combined p-value for the global hypothesis is 0.000127. Higgins and Noble (1993) conclude that decay of treatment B is faster than that of A, even though at the last time point, $T = 5$, substantially the same distribution of pH values is observed ($p = 0.068$).

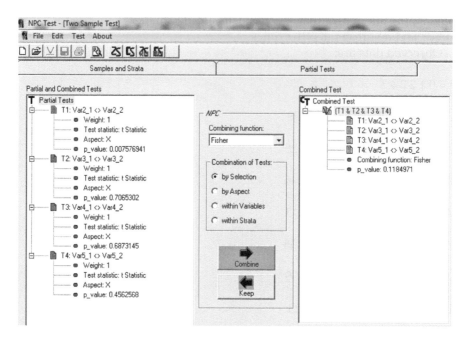

FIGURE 5.2
Use of the NPC test to compare two samples via Pesarin's method.

Pesarin's Nonparametric Combination

The methods described in this section have the advantage that they apply to
continuous, ordinal, or categorical variables or to any combination thereof.
They can be applied to one-, two-, or k-sample comparisons. As in the preceding section, suppose we have made a series of exchangeable vector-valued
observations on K subjects, each vector consisting of the values of J variables.
The first variable might be a 0 or 1, according to whether or not the kth seedling in the ith treatment group germinated; the second might be the height
of the kth seedling; the third the weight of its fruit; the fourth a subjective
appraisal of fruit quality; and so forth. With each variable is associated a
specific type of hypothesis test, the type depending on whether the observation is continuous, ordinal, or categorical, and whether it is known to have
come from a distribution of specific form. Let **To** denote the vector of single-
variable test statistics derived from the original unpermuted matrix of observations. These might include differences of means or weighted sums of the
total number germinated, or any other statistic one might employ when testing just one variable at a time. When we rearrange the treatment labels on the
observation vectors we obtain a new vector of single-variable test statistics:
$T\pi$.

In order to combine the various tests, we need to reduce them all to a common scale. Proceed as follows:

1. Generate S permutations of X, and thus obtain S vectors of univariate test statistics.

2. Rank each of the single-variable test statistics separately. Let R_{ij} denote the rank of the test statistic for the jth variable for the ith permutation when it is compared to the value of the test statistic for the jth variable for the other S permutations, including the values for the original test statistics.

3. Combine the ranks of the various univariate tests for each permutation using Fisher's omnibus method:

$$U_i = -\sum_{j=1}^{J} \log\left[\frac{S + 0.5 - R_{ij}}{S+1}\right]; i = 1, \ldots S$$

4. Reject the null hypothesis only if the value of U for the original observations is an extreme value.

This straightforward yet powerful method is due to Pesarin (1990). A SAS macro to perform Pesarin's omnibus test may be downloaded from http://homes.stat.unipd.it/pesarin/NPC.SAS.

Summary

In this chapter, you learned two separate approaches to the multivariate analysis of data.

You applied a single-value test statistic to a two-sample comparison and to the analysis of repeated measures. You learned how to detect clustering in time and space and to validate clustering models. You used the generalized quadratic form in its several guises, including Mantel's U and Mielke's multiresponse permutation procedure (MRPP), to work through applications in archeology, epidemiology, and ornithology.

You learned two methods based on Fisher's nonparametric combination rule for combining the results from several univariate tests.

To Learn More

Blair et al. (1994), Mielke and Berry (1999), and van-Putten (1987) review alternatives to Hotelling's T^2. Extensions to other experimental designs

are studied by Pesarin (1997, 2001) and Barton and David (1961). For
additional insight into the analysis of repeated measures, see Zerbe and
Murphy (1986).

Mantel's U has been rediscovered frequently, often without proper attribu-
tion (see Whaley, 1983). Empirical power comparisons between MRPP rank
tests and with other rank tests are made by Tracy and Tajuddin (1986) and
Tracy and Khan (1990).

Potter and Griffiths (2006) compare various combining functions to be used
in conjunction with Pesarin's permutation method and describe a number of
applications. Hayasaka and Nichols (2004) use Pesarin's combining function
to analyze brain image data. Giancristofaro and Bonnini (2009) show how it
may be applied to ordinal categorical variables.

Exercises

1. Using the data from http://www.statcrunch.com/5.0/index.php?
   dataid=376459, determine whether men have significantly higher
   stat scores.

2. Using the data from http://www.statcrunch.com/5.0/index.php?
   dataid=401730, determine which league fields the strongest teams.

3. Show that the Pearson correlation coefficient is an example of
   Mantel's U statistic.

4. Sixteen seedlings were planted in two trays, eight per tray. The first
   tray contained a new fertilizer but was otherwise the same as the
   second tray. The heights of the surviving plants were compared after
   two weeks:

   Tray 1: 5, 4, 7, 7.5, 6, and 8 in.

   Tray 2: 5, 4.5, 3, and 6 in.

   a. Can this be viewed as a multivariate problem?

   b. If so, use Pesarin's test to analyze the results.

5. Are there significant differences in the financial ratios of solvent and
   insolvent firms? Use the part of the data collected by Trieschman
   and Pinches (1973) reproduced below to make your determination.
   Apply both Hotelling's T^2 and Pesarin's test.

Insolvent	Agents' Balances/ Total Assets	Stocks—Cost/ Stocks—mktvlu	Bonds—Cost/ Bonds—mktvlu	Expenses Paid/ Premiums
0	0.06	0.4	1.14	0.11
0	0.06	0.76	1.01	0.09
0	0.03	0.85	1	0.12
0	0.03	0.9	1	0.06
0	0.05	0.93	1.21	0.19
0	0.06	1.58	1.01	0.15
0	0.02	0.38	1	0.14
0	0.08	0.98	1	0.19
1	0.06	1.17	1.15	0.73
1	0.05	0.7	1.05	0.05
1	0.17	0.85	1	0.09
1	0.06	0.59	0	0.06
1	0.34	0.9	1.03	0.01
1	0.23	1	1.16	0.01
1	0.11	0.93	0.98	0.25
1	0.19	1.12	1.06	0.5

6

Experimental Design and Analysis

The purpose of an experimental design is to eliminate or at least reduce the effects of extraneous factors so as to focus attention on the factors that are of primary interest and increase the power of our statistical tests. The advantage of resampling methods is that they allow us to analyze an experiment in the way it was designed.

We encountered an experimental design, matched pairs, in the "Test for Nonequivalence" section. By eliminating the variation from individual to individual, we increased the power of our tests to detect deviations from the null hypothesis.

In this chapter, we design experiments to compare multiple treatments simultaneously, to test separately the effects of various factors, and to do so in a way that will ensure the various tests are independent of one another, to correct for carryover effects when treatments are administered sequentially, and to compensate for missing data. In each instance, we must first address two concerns:

1. What test statistic should we use?
2. Which sets should we permute?

Separating Signal from Noise

All the problems we examine in this text would be straightforward if it weren't for the noise in the form of subject-to-subject variation that nature has imposed on the relationships that we're trying to make sense of. In statistical terms, the effect we are interested in is *confounded* with many effects with which we are not. Fortunately, in many cases, but not all, properly applied experimental design can reduce the noise and sharpen the picture. It also can help us avoid mistaking noise for signal.

Eliminating or reducing extraneous variation is the first of several preventive measures we use each time we design an experiment or survey. We strive to conduct our experiments in a biosphere with atmosphere and environment totally under our control. And when we can't, which is almost always the case, we record the values of the extraneous variables to use them either as blocking units or as covariates.

Blocking

Although the significance level of a permutation test may be distribution-free, its power, that is, its ability to detect deviations from the null distribution, depends on the underlying distribution.* The more variable our observations, the more difficult it becomes to detect the alternative. With widgets coming off an assembly line, we can get around the problem simply by increasing the sample size. But humans are far more heterogeneous. They come in all shapes and sizes, races, and ages. Some smoke, some don't. Some garden plots like that in my backyard consist of nonporous infertile clay, while others across the street boast of a rich sandy loam. One way to reduce the variance of our experiments and surveys is to *block* them, that is, to subdivide the population into more homogeneous subpopulations and to take separate independent samples from each.

The basis for blocking lies in the potential sources of variation. These usually can be grouped as follows:

- Characteristics of the experimental subject, such as age and sex
- Characteristics of the environment, such as temperature and humidity
- Characteristics of the observer and the observing methodology

Suppose you were designing a survey on the effect of income level on the respondent's attitude toward health insurance reform. The views of men and women may differ markedly on this controversial topic. To reduce the variance and increase the power of your tests, block the experiment, interviewing and reporting on men and women separately. You may also want to adjust for different ages and different levels of income.

Whenever a population can be subdivided into distinguishable subpopulations, you can reduce the variance of your observations and increase the power of your statistical tests by blocking or stratifying your sample.

Suppose we have agreed to divide our sample by gender into two blocks. If this were an experiment, rather than a survey, we would then assign subjects to treatments separately and independently within each block. Suppose we have 10 experimental subjects, 4 men and 6 women. Half of each group are to receive the standard treatment and half the new experimental one. We could assign the men in any of 4 choose 2 or 6 ways and the women in any of 6 choose 3 or 20 ways, for a total of $6 \times 20 = 120$ possible random assignments in all.

When we come to analyze the results of our experiment, we use the permutation approach to ensure that we analyze in the way the experiment was designed. Our test statistic is a natural extension of that used for the two-sample case: the sum of the sums of the observations in the first sample of each block. In symbols,

---

* If it's been a long time since your last statistics course, we've included a glossary of statistical terms as an appendix.

$$\sum_{j=1}^{blocks} \sum_{i=1}^{n[1]} X_{j1i}$$

where Xj_1i is the ith observation in the first sample in the jth block.

The advantage of using resampling methods is that the experiment need not be balanced. Note that our test statistic is a sum of several independent sums. Unequal sample sizes resulting from missing data or an inability to complete one or more portions of the experiment will affect the analysis only in the relative weights assigned to each subgroup.

Analyzing a Blocked Experiment

We consider two quite different cases in this section:

1. Combining data to obtain improved estimates
2. Comparing samples from two populations

Combining Data to Obtain Improved Estimates

By combining data from several sources, we can increase the precision of our estimates. In the following R listing, the weight given to each sample is inversely proportional to its standard deviation.

```
#Combine and weight samples to obtain a CI for population median
first=c(3.87, 4.48, 5.00, 3.60, 3.89, 2.50, 4.72, 3.15, 2.94,
4.65, 4.59, 3.60, 3.67, 3.16)
second=c(8.67, 2.64, 4.01, 4.31, 2.75, 2.44, 2.95, 7.86, 3.36,
6.97, 1.97, 5.31)
#record group sizes
n1 = length(first)
n2=length(second)
n=n1+n2
#record standard deviations
s1=sqrt(var(first))
s2=sqrt(var(second))
ratio= (n1/s1)/(n1/s1 + n2/s2)
#set number of bootstrap samples
N =100
stat = numeric(N) #create a vector in which to store the results
#the elements of the vector will be numbered from 1 to N
#Set up a loop to generate a series of bootstrap samples
for (i in 1:N){
   choice = sort(runif (n,0,1))
   cnt=0
   while (choice[cnt+1]<ratio) cnt=cnt+1
   #bootstrap sample counterparts to observed samples are
denoted with "B"
```

```
firstB= sample (first, cnt, replace=T)
secondB=sample(second,n-cnt,replace=T)
stat[i] = median(c(firstB,secondB))
}
quantile(stat,c(.05,.95))
```

Comparing Blocked Samples from Two Populations

For C++, R, and resampling methods, comparing samples from two populations is merely a question of embedding a second loop inside the original
loop of our program code for the two-sample comparison. For example, in R:

```
for (k in 1:MC){
stat=0
for (j in 1:blocks){
D= sample (A[j],n)
stat=stat+sum(D)
}
if (stat <= sumorig)cnt=cnt+1
}
(cnt+1)/(MC+1) #pvalue
```

With Stata, one needs to enter the blocking variable along with the other
data, then make use of the stratified permute command as follows:

```
block      regimen      weight loss

man        diet              50
man        diet              25
man        diet              30
man        exercise          22
man        exercise          34
man        exercise          28
woman      diet              30
woman      diet              20
woman      exercise          28
woman      exercise          25
woman      exercise          24
```

```
. encode regimen, gen(treat)
. permute wtloss "sum wtloss if treat==1" sum=r(sum),
reps(400) strata(block) nowarn

command: sum wtloss if treat==1
statistic: sum = r(sum)
permute var: wtloss

Monte Carlo permutation statistics Number of obs = 11
Number of strata = 2
Replications = 400
```

```
T T(obs) c n p=c/n SE(p) [95% Conf. Interval]
sum 155 117 400 0.2925 0.0227 .248349 .3397525
```

```
Note: confidence interval is with respect to p=c/n
Note: c = #{T >= T(obs)}
```

PRINCIPLES OF EXPERIMENTAL DESIGN

List factors you feel may influence the outcome of your experiment; that is, list all potential sources of variation.

Block factors under your control. You may want to use some of these factors to restrict the scope of your experiment; e.g., eliminate all individuals under 18 and over 60.

Measure remaining factors.

Randomly assign units to treatment within each block.

Randomization

To illustrate these design principles, let's set up an experiment to find a new and better fertilizer for growing tomatoes. Ideally, we would perform this experiment indoors under artificial lights in a climate-controlled chamber so that we would have total control over the amounts of daylight and moisture and of the temperature. To assess the effect of soil type on growth at the same time, we'll plant our seeds in three separate trays containing sandy loam, a sand-clay mix, and a clay soil, respectively.

A word of caution: Studies have shown there is a natural tendency to plant the tall seedlings first and save the runts for last. To avoid confounding height with treatment, we could deal the seedlings out like cards, the first for tray 1, the second for tray 2, and so forth down the line. This method would still seem to put the best plants into the first tray.

In order to analyze an experiment using permutation methods, we need to be sure that each assignment of subjects to treatment is equally likely. Only if we assign the plants to the different trays completely at random will this assumption be fulfilled. If the plants assigned to tray 1 prove in the end to be larger on the average than those in tray 2, this should be strictly a matter of chance, not the result of a specific bias on our part.

Randomly assign subjects to treatment whenever you can't control all the factors in an experiment.

Our tomato study might use either of the following randomization schemes:

1. Rearrange the integers 1 to *m* at random, where *m* is the number of trays. Place the first *m* plants in the trays in the order specified. Choose a second random rearrangement, a third, and so on, until all the plants have been placed in the trays.

2. Throw an m-sided die and place the first plant in the indicated tray. Continue to throw the die until all the plants have been placed. If the die indicates that an already full tray has been selected, throw again. (See also Exercise 4.)

k-Sample Comparison

When samples from more than two populations are under consideration, the statistic we use will depend upon both the alternatives we wish to test and the losses we will incur if a type I or type II error is made.

Testing for Any and All Differences among Means

Suppose we are trying to choose one of three brands of fertilizer to use on our company's fields. Prices vary, and if there are no real differences in the crop yield resulting from the use of one fertilizer rather another, we'd just as soon purchase the cheapest brand. We speculate that crop yield is a function of the fertilizer used and a random component that is the cumulative result of a large number of factors that we simply cannot control, such as minor differences in soil pH and so forth. In symbols, our *additive* model is that

$$X_{ij} = \mu + f_i + z_{ij}$$

where X_{ij} is the yield of the jth planting treated with the ith fertilizer, f_i is the effect of the ith fertilizer, $\Sigma f_i = 0$, and the experimental errors $\{z_{ij}\}$ are independent of one another and all come from the same probability distribution whose mean is zero. Note that the expected value of X_{ij} is $\mu + f_i$.

The sum of squares of the deviations about the grand mean may be analyzed into two sums:

$$\sum_{i=1}^{I}\sum_{j=1}^{n_j}(X_{ij} - \bar{X}_{..})^2 = \sum_{i=1}^{I}\sum_{j=1}^{n_j}(X_{ij} - \bar{X}_{i.})^2 + \sum_{i=1}^{I}n_i(\bar{X}_{i.} - \bar{X}_{..})^2$$

the first of which represents the within-treatment sum of squares and the second the between-treatment sum of squares.

For testing the null hypothesis that the effects of each fertilizer are the same, against the alternative that one of the fertilizers is superior to the others, only the between-treatment sum is of interest. Examining this sum for factors that are the same for all permutations (and thus would be a waste of time to calculate), we may neglect the grand mean. Only the sum

$$F_2 = \sum_{i=1}^{I} n_i (\bar{X}_{i.})^2 = \sum_{i=1}^{I} \left(\sum_{j=1}^{n_i} X_{ij} \right)^2 \Big/ n_i$$

changes as we rearrange the observations among levels, and it is this sum whose permutation distribution we use to test the hypothesis.

In simulations conducted by Good and Lunneborg (2006), this permutation test was found to be more powerful than the standard parametric test if there were unequal numbers in the various treatment groups, and as powerful otherwise.

F_2 emphasizes large deviations from the mean; if we wish to attribute equal value to all deviations, we should use as our test statistic

$$F_1 = \sum_{i=1}^{I} n_i \mid \bar{X}_{i.} - \bar{X}_{..} \mid .$$

We may also want to substitute medians for means, or replace the original observations by their ranks or some other transformation as described in Chapter 5. If the changes are more likely to be percentages of some standard value, that is, if our model is that $X_{ij} = \mu f_i z_{ij}$, then we must first take the logarithms of our observations, so that $\log(X_{ij}) = \eta + gi + \varepsilon_{ij}$.

Testing for Any and All Differences among Variances

The test for comparing the variances of two samples based upon the absolute deviations about the sample medians (see the "Preferred Method" section) is easily generalized to k-samples from k-populations. Such a test would be of value, for example, to test for the equality of variances prior to performing a k-sample analysis for a difference in means among test groups.

First, we create k sets of deviations about the sample medians and make use of the test statistic

$$S = \sum_{k=1}^{K} \left(\sum_{j=1}^{n_k^{-1}} x'_{kj} \right)^2$$

The choice of the square of the inner sum ensures that this statistic takes its largest value when the largest deviations are all together in one sample after relabeling.

To generate the permutation distribution of S, we again have two choices. We may consider all possible rearrangements of the sample labels over the k sets of deviations. Or, if the samples are equal in size, we may first order the deviations within each sample, group them according to rank, and then rearrange the labels within each ranking

Again, this latter method is directly applicable only if the *k*-samples are equal in size, and again, this is unlikely to occur in practice. We will have to determine a confidence interval for the *p*-value for the second method via a bootstrap in which we first select samples from samples (without replacement) so that all samples are equal in size.

R

```
F1=function(size,data){
   #size is a vector containing the sample sizes
   #data is a vector containing all the data in the same order
as the sample sizes
   stat=0
   start=1
   grandMean = mean(data)
   for (i in 1:length(size)){
      groupMean = mean(data[seq(from = start, length =
size[i])])
      stat = stat + abs(groupMean-grandMean)
      start = start + size[i]
   }
   return(stat)
}
```

We use this function repeatedly in the following R program.

Suppose we wish to compare the effects of three brands of fertilizer on crop yield recorded as follows:

```
FastGro = c(27, 30, 55, 71, 18)
NewGro = c(38, 12, 72)
WunderGro = c(75,76,54)

# One-way analysis of unordered data via a Monte Carlo
size = c(4,6,5,4)
data = c(FastGro, NewGro, WunderGro)
f1=F1(size,data)
#Number MC of simulations determines precision of p-value
MC = 1600
cnt = 0
for (i in 1:MC){
   pdata = sample(data)
   f1p=F1(size,pdata)
   # counting number of rearrangements for which F1 greater
than or equal to original
   if (f1 <= f1p) cnt=cnt+1
}
(cnt+1)/(N+1)
```

Stata

```
*** program to calculate F2 due to Lynn Markham
capture program drop f2test
program define f2test, rclass
tempvar xijsq sumxj obsnum f2 bcount
sort brand
qui by brand: gen `obsnum'=_n
qui by brand: gen `bcount'=_N
egen `sumxj'=sum(growth), by(brand)
replace `sumxj'=. if `obsnum'>1
gen `xijsq'=(`sumxj'*`sumxj')/`bcount'
egen `f2'=sum(`xijsq')
return scalar F2=`f2'[1]
end
```

Ordered Alternatives

Suppose we want to assess the effect on crop yield of hours of sunlight, observing the yield for I different levels of sunlight, with ni observations at each level. The null hypothesis would be that sunlight has no effect on production, while our ordered alternative would be that yield is an increasing function of sunlight.

While the statistics F_1 and F_2 offer protection against a broad variety of alternatives, they do not provide a most powerful test against the ordered alternative that is our specific interest.

Suppose we suspect that yield is a specific function $f[d]$ of the number of days d of sunlight. If we have measured crop yield on land that received d_1, d_2, ..., dI days of sunlight, the optimal statistic for distinguishing between the null hypothesis and our ordered alternative would be the Pearson correlation

$$\sum_{i=1}^{I} f[d_i] \sum_{j=1}^{n_i} y_{ij} \, ,$$

where yij is the yield of the jth plot receiving the ith level of sunlight.

It is the optimal choice for test statistic because while all values are equally likely under the null hypothesis, large values are more likely if the ordered alternative is true.

Frank et al. (1978) studied the increase in chromosome abnormalities and micronuclei as the dose of various known mutagens was increased. Their object was to develop an inexpensive but sensitive biochemical test for mutagenicity that would be able to detect even marginal effects. Thus, they were more than willing to trade the global protection offered by the F-test for a statistical test that would be sensitive to ordered alternatives.

TABLE 6.1

Micronuclei in Polychromatophilic Erythrocytes
and Chromosome Alterations in Bone Marrow
of CY-Treated Mice

Dose (mg/kg)	Number of Animals	Micronuclei per 200 Cells	Breaks per 25 Cells
0	4	0 0 0 0	0 1 1 2
5	5	1 1 1 4 5	0 1 2 3 5
20	4	0 0 0 4	3 5 7 7
80	5	2 3 5 11 20	6 7 8 9 9

In the data collected by Frank et al. shown in Table 6.1, the anticipated effect is proportional to the logarithm of the dose, so we take $f[\text{dose}] = \log[\text{dose} + 1]$. (Adding a 1 to the dose keeps this function from blowing up at a dose of zero.)

They studied four dose groups; the original data for breaks may be written in the form

$$0\ 1\ 1\ 2\ 0\ 1\ 2\ 3\ 5\ 3\ 5\ 7\ 7\ 6\ 7\ 8\ 9\ 9$$

Our test statistic is the Pearson correlation, which Good (2009) has shown is highly robust against deviations from normality. In this example, it takes the value of 0.82, and a statistically significant ordered dose response ($\alpha < 0.001$) has been detected. The micronuclei also exhibit a statistically significantly result when we calculate the correlation of S with $f[i] = \log[\text{dose}_i + 1]$.

A word of caution: If we use some function of the dose other than $f[\text{dose}] = \log[\text{dose} + 1]$, we might not observe a statistically significant result. Our choice of a test statistic must always make practical as well as statistical sense.

Multiple Factors

Suppose we want to assess the simultaneous effects on crop yield of both hours of sunlight and rainfall. We determine to observe the crop yield X_{ijm} for I different levels of sunlight, $i = 1$ to I, and J different levels of rainfall, $j = 1$ to J, and to make M observations at each factor combination ij, $m = 1$ to M. We adopt as our model relating the dependent variable crop yield (the effect) to the independent variables of sunlight and rainfall (the causes):

$$X_{ijm} = \mu + si + rj + (sr)ij + \varepsilon ijm$$

where $\Sigma s_i = 0$, $\Sigma rj = 0$, $\Sigma i(sr)_{ij} = 0$, and $\Sigma j(sr)_{ij} = 0$.

In this model, terms with a single subscript like s_i, the effect of sunlight, are called *main effects*. Terms with multiple subscripts like $(sr)_{ij}$, the residual and nonadditive effect of sunlight and rainfall, are called *interactions*. The residuals $\varepsilon_{ij}m$ represent that portion of crop yield that cannot be explained by sunlight and rainfall alone. To ensure the residuals are exchangeable so that permutation methods can be applied, the experimental units must be assigned at random to treatment.

When we have multiple factors, we must also have multiple test statistics. In the preceding example, we require two separate tests and test statistics for the main effects of rainfall and sunlight, plus a test for their interaction. Will we be able to find statistics that measure a single intended effect without confounding it with a second unrelated effect? Will the several test statistics be independent of one another?

The answer is yes to both questions only if the design is *balanced*, that is, if there are equal numbers of observations in each subcategory. Moreover, only symmetric permutations (defined in Chapter 11) will ensure the test statistics are independent of one another. In an unbalanced design, main effects will be confounded with interactions, so the two can not be tested separately, a topic we return to with a bootstrap solution in the "Determining Sample Size" section.

Main Effects

We can assess the main effects using essentially the same statistics we would use for randomized blocks.

If we have only two levels of sunlight, then our test statistic for the effect of sunlight was shown in Chapter 4 to be the sum of all observations at the first level of sunlight.

If we have more than two levels of sunlight, our test statistic may be any of the three test statistics described in the "*k*-Sample Comparison" section, the distinction being that we need to sum over the blocks for the various levels of rainfall, as in the "Blocking" section. For example, to compute the correlation for the main effect of sunlight, we would use the statistic

$$\sum_{j=1}^{J} \sum_{i=1}^{I} f[d_i] \sum_{m=1}^{n_{ij}} y_{ijm}$$

To obtain the permutation distributions of the test statistics for the effect of sunlight, we permute the observations separately and independently in each of the J blocks determined by a specific level of rainfall.

Let's apply the permutation method to determine the main effects of sunlight and fertilizer on crop yield using the data from the two-factor experiment depicted in Table 6.2. As there are only two levels of sunlight in this

TABLE 6.2

Effect of Sunlight and Fertilizer on Crop Yield

| | | Fertilizer | |
Sunlight	Low	Medium	High
Low	5	15	21
	10	22	29
	8	18	25
High	6	25	55
	9	32	60
	12	40	48

TABLE 6.3

Crop Yield Data Rearranged

| | | Fertilizer | |
Sunlight	Low	Medium	High
Low	6*	15	21
	10#	22	29
	8	18	25
High	5*	25	55
	9#	32	60
	12	40	48

experiment, after eliminating terms that are invariant under rearrangement of the labels, we use the sum of the sums of the observations in the first sample of each block to test for the main effect.

For the original observations, this sum is 23 + 55 + 75 =153. One possible rearrangement is shown in Table 6.3, in which we have interchanged the two observations marked with an asterisk, the 5 and 6. The new value of our test statistic is 154.

As can be seen by a continuing series of straightforward hand calculations, the test statistic for the main effect of sunlight is as small or smaller than it is for the original observations in only 8 out of the $\binom{6}{3}^3 = 8{,}000$ possible rearrangements. For example, it is smaller when we swap the 9 of the high-low group for the 10 of the low-low group (the two observations marked with the pound sign # in Table 6.3). The effect of sunlight is statistically significant.

The computations for the main effect of fertilizer are more complicated— we must examine $\binom{9}{3\ 3}^2$ rearrangements, and compute the statistic F_1 for each. We use F_1 instead of the correlation because of the possibility that too much fertilizer—the "high" level—might actually suppress growth. Only a computer can do this many calculations quickly and correctly, so we adapted one of the programs in the "Testing for Any and All Differences among Means" section to make them. The estimated significance level is 0.001, and we conclude that this main effect, too, is statistically significant.

In this last example, each category held the same number of experimental subjects. If the numbers of observations were unequal, our main effect would have been confounded with one or more of the interactions (see Section 5.6). In contrast to the simpler designs we studied in Chapter 4, missing data will affect our analysis.

Testing for Interactions

In the preceding analysis of main effects, we assumed the effect of sunlight was the same regardless of the levels of the other factors. But this may not always be the case. Of what value is sunlight to a plant if there is not enough fertilizer in the ground to support growth? And vice versa, of what value is fertilizer if there is insufficient sunlight?

Sunlight and fertilizer interact; we cannot simply add their effects. Our model should and does include an interaction term $(sf)_{ij}$:

$$X_{ijm} = \mu + s_i + f_j + (sf)_{ij} + \varepsilon_{ij}m$$

The presence of the constant term μ allows us to simplify the analysis by setting $\Sigma_i s_i = \Sigma_j f_j = \Sigma_i (sf)_{ij} = \Sigma_j (sf)_{ij} = 0$, so that the various subscripted terms represent deviations from an overall average.

Suppose we were to eliminate row and column effects by subtracting the row and column means from the original observations, as in Table 6.4.

$$X'_{ijm} = X_{ijm} - \bar{X}_{i..} - \bar{X}_{.j.} + \bar{X}_{...}$$

TABLE 6.4

Residuals after Eliminating Main Effects

Sunlight	Fertilizer		
	Low	Medium	High
Low	4.1	–2.1	–11.2
	9.1	4.1	–3.2
	7.1	0.1	–7.2
High	–9.8	–7.7	7.8
	–7.8	–0.7	12.8
	–3.8	7.2	0.8

The pattern of plus and minus signs in this table of residuals suggests that fertilizer and sunlight affect crop yield in a superadditive or synergistic fashion. Note the minus signs associated with the mismatched combinations of a high level of sunlight and a low level of fertilizer, and a low level of sunlight with a high level of fertilizer.

An obvious statistic to test the hypothesis that the interactions terms are zero for all levels of sunlight and fertilizer would appear to be $W_{ij} = \Sigma_i \Sigma_j (\Sigma m X'_{ij} m)^2$.

Unfortunately, the labels on the deviates X'_{ijm} are not exchangeable, so we cannot permute them. The expected value of X'_{ijm} is $\varepsilon_{ijk} - \overline{\varepsilon}_{i..} - \overline{\varepsilon}_{.j.} + \overline{\varepsilon}_{...}$. Thus, the deviates are correlated and the correlation between two deviates in the same row or column has a different value than the correlation between deviates in different rows or columns.

We consider a possible solution to this problem in Chapter 11 through the use of synchronized rearrangements.

Eliminating the Effects of Multiple Covariates

Similar problems arise should we attempt to eliminate the effects of multiple covariates by permuting the residuals. To circumvent such difficulties, Tang et al. (2009) propose a permutation test that permutes treatment assignments for entire observations. Their technique mirrors the design of the study, including adjustments for nonresponders and for baseline covariates. Specifically, they permute the randomization indicators according to the blocked randomization design, and then evaluate the intervention effect following the procedures in the original data analysis protocol, including weighting and covariate adjustments.

Consider a regression function: $E(Y) = g(\beta_0 + \beta_1 X + \beta_2 Z)$, where Z is the treatment indicator (1 = intervention, 0 = control), X is the vector of covariates, and g denotes a link function, with the identity link function appropriate for a linear model and the inverse of the logit function appropriate for a logistic regression model. The vector $\mathbf{X}$ could include fixed effects for blocks, or one can proceed with separate block effects.

The intervention effect is given by the regression coefficient β_2; the significance of the intervention effect is tested with the null hypothesis H_0: $\beta_2 = 0$ vs. H_a: $\beta_2 \neq 0$. The ratio $T = \beta_2/se[\beta_2]$ is used as the test statistic.

The null distribution of T for testing the hypothesis H_0 is simulated by rerandomizing the treatment assignment according to the original randomization protocol while keeping outcomes and covariates as observed. For a blocked randomized study with attrition weighting, follow this three-step procedure:

Step 1: Compute the test statistic for the actual data observed:

a. Select an attrition weighting model.

b. Fit the attrition weighting model.

c. Derive attrition weights using the model fitted in step 1b.

d. Fit the regression model using linear or logistic regression to obtain the test statistic T for the null hypothesis.

Step 2: Estimate the null distribution for the test statistic T with N replicates of the three substeps below:

a. Rerandomize treatment assignment within each block.

b. Rederive the attrition weights using the methods in step 1a–c.

c. Refit the regression model using SUDAAN for linear or logistic regression with attrition weights rederived in step 2b to rederive the test statistic T.

Step 3: Derive the empirical p-value for the test statistic T from the null distribution based on the permutation distribution obtained in step 2, i.e., $p = (M + 1)/(N + 1)$, where M denotes the number of replicates for which the test statistic T obtained in the permutation procedure in step 2 is equal to or greater than (in absolute value) the observed value of T obtained in step 1, and N denotes the total number of replicates. The addition of 1 in both numerator and denominator represents the observed test statistic for the original data, which must be considered as one of the realizations of the permutation distribution.

Latin Squares

While it might be convenient to fertilize our plots as shown in Table 6.5, the result could be a systematic bias. For example, suppose there is a gradient in dissolved minerals from east to west across the field.

To avoid biasing our results, we should assign the fertilizer to plots at random, thus ensuring the exchangeability of the error terms and the exactness of the corresponding permutation test. The layout adopted in Table 6.6,

TABLE 6.5

Systematic Assignment of Fertilizer Levels to Plots

W		E
High	Medium	Low
High	Medium	Low
High	Medium	Low

TABLE 6.6

Random Assignment of Fertilizer
Levels to Plots

High	Medium	Low
Low	Medium	Low
High	High	Medium

TABLE 6.7

Latin Square Assignment
of Fertilizer Levels to Plots

High	Medium	Low
Low	High	Medium
Medium	Low	High

obtained with the aid of a computerized random number generator, is one
example. Alas, while this layout reduces the effects of this hypothetical gradient, it does not eliminate them.

We can avoid such an undesirable event by making use of a *Latin square*
(Table 6.7) in which each fertilizer level occurs once and once only in each
row and in each column. This layout ensures that any systematic gradient of
minerals in the soil will have almost equal impact on each of the three treatment levels. It will have an almost equal impact even if the gradient extends
from northeast to southwest rather than from east to west or north to south.

The Latin square is one of the simplest examples of an experimental design
in which the statistician is able to reduce the overall sample size. A Latin
square is a three-factor experiment in which each combination of factors
occurs once and once only. We can use a Latin square like that of Table 6.7 to
assess the effects of soil composition on crop yield:

Suppose that factor 1, gypsum concentration, say, is increasing from left to
right, and factor 2 is increasing from top to bottom (or from north to south).
Note that the third factor, treatment, occurs in combination with the other
two in such a way that each combination of factors—row, column, and treatment—occurs once and once only.

Because of this latter restriction, there are only 12 different ways in which
we can assign the varying factor levels to form a 3 × 3 Latin square. Among
the other 11 designs are the following, where the varying levels of the third
factor are denoted by the capital letters A, B, and C,

Design 2

	1	2	3
1	A	C	B
2	B	A	C
3	C	B	A

and

Design 3

```
    1  2  3
1   C  B  A
2   B  A  C
3   A  C  B
```

Assume we begin our experiment by selecting 1 of these 12 designs at random and planting our seeds in accordance with the indicated conditions.

Because there is only a single replication of each factor combination in a Latin square, we cannot estimate the interactions. The Latin square is appropriate only if we feel confident in assuming the effects of the various factors are completely *additive*; that is, the factors have neither antagonistic nor synergistic effects.

Our model for the Latin square is

$$X_{kji} = \mu + q_k + r_j + s_i + \varepsilon_{kji}$$

with the effects as always representing deviations from an average, so that the sums of the main effects are zero. As always in a permutation analysis, we assume the labels on the errors $\{\varepsilon_{kji}\}$ are exchangeable. Our null hypothesis is that the additive effects of the various levels of sunlight $\{s_i\}$ are all zero. If we assume an ordered alternative, $K: s_1 < s_2 < s_3$, our test statistic for the main effect is similar to the correlation statistic

$$\sum_{i=1}^{3} (i-1)(\bar{X}_{i..} - \bar{X}_{...})$$

or, equivalently, after eliminating the grand mean $\bar{X}_{...}$, which is invariant under permutations, $R' = \bar{X}_A - \bar{X}_C$.

We evaluate this test statistic both for the observed design and for each of the 12 possible Latin square designs that might have been employed in this particular experiment. We reject the hypothesis of no treatment effect only if the test statistic R for the original observations is an extreme value.

For example, suppose we employ the Latin square of Table 6.7 and observe

21 28 17

14 27 19

13 18 23

Then $\bar{X}_A - \bar{X}_C = 58 - 57 = 1$. Had we employed design 2, then $R' = 71 - 65 = 6$, while with design 3, our test statistic $57 - 58 = -1$.

We see from the permutation distribution obtained in this manner that 1, the value of our test statistic for the design actually employed in the experiment, is an average value, not an extreme one. We accept the null hypothesis and conclude that increasing the treatment level from A to B to C does not significantly increase the yield.

DESIGNING AND ANALYZING A LATIN SQUARE

Null hypothesis: Mean/median the same for all levels of each treatment.
Alternative hypothesis: Means/medians are different for at least one level.

Assumptions:

1. Observations are exchangeable if the hypothesis is true.
2. Treatment effects are additive (not synergistic or antagonistic).

Procedure:

List all possible Latin squares for the given number of treatment levels.
Assign one design at random and use it to perform the experiment.
Choose a test statistic (R, F_1, or F_2).
Compute the statistic for the design you used.
Compute the test statistic for all the other possible Latin square designs.
Determine from the resultant permutation distribution whether the original value of the test statistic is an extreme one.

Crossover Designs

A crossover design should be used in comparisons of treatments when each subject acts as his or her own control and there is a possibility of a carryover of the effect of one of the treatments into the time period devoted to the next. A 2 × 2 crossover design would take the following form:

	Time Period	
	1	2
Treat A	Group a	Group b
Treat B	Group b	Group a

That is, a different group of subjects receives the treatment in each successive time period. If no data are missing at the end of the trials, then we are in a position to apply the two-sample comparison of means studied in Chapter 4, exchanging labels between those subjects who received the control treatment and those who did not (Guilbaud, 1999; Johnson and Mercante, 1996; Patefield, 2000). But if subjects drop out of the trials or if treatments are administered over more than two time periods, such exchanges of treatment labels would confound carryover with main effects.

Analysis of a Complete Balanced Design

An example of such a design is one in which patients were assigned at random to one of the three possible treatment sequences:

$$TTC \ TCT \ CTT$$

$$TCT \ CTT \ TTC$$

$$CTT \ TTC \ TCT$$

where a T indicates that the patient received the new treatment on the day in question, and C that he or she received a placebo control. As shown above, for each episode in the sequence, two-thirds of the patients receive the active treatment. The three "super blocks" (TTC, TCT, and CTT) form a Latin square.

Our test statistic is the sum of the observations with the control label, $S = \sum\sum\sum C_{ijk}$, or simply $\sum C$, where i denotes the treatment sequence, j the period, and k the subject. Other test statistics are possible, but those that include summations that are invariant under permutations of the labels such as the sum of all the observations ($\sum T + \sum C$), the within-group sum of squares that forms the denominator of the F-ratio, and the difference of the two group means $2\sum C - \sum T = 3\sum C - (\sum T + \sum C)$ are equivalent to the one proposed here and thus are unnecessarily complex.

The permutation distribution is obtained by rearranging the *sequence labels* on the subjects so that the number of subjects $\{n_i\}$ in each sequence is preserved. In the crossover study of the effects of fentanyl buccal tablets for relief of breakthrough pain in opioid-treated patients with chronic low back pain analyzed by Good and Xie (2008), different numbers of patients $\{ni, i = 1, 2, 3\}$ were assigned to each of the treatment sequences. But as the exchange of sequence labels resulted only in the exchange of observations within the periods, differences among the periods did not affect the permutation distribution. If the treatment has no effect, and all observations have equal inherent variance, then all values of the permutation distribution are equally likely. On the other hand, if the treatment results in a decrease in the expected value of an observation, then one would expect the original value of the test statistic to be at the high end of the permutation distribution.

Analysis of a Balanced Design When Not All Subjects Complete Treatment

Although 74 patients entered the fentanyl study, only 61 patients completed all nine periods, a not unexpected result in a clinical trial. Still, nine of the remaining patients completed at least the first six periods. As the first six periods also constitute a balanced design, Good and Xie (1988) proceeded to make use of the data from the patients who did not complete all nine periods, as follows.

All the data from patients who completed all nine periods were retained as a single block. A second block was formed from the data for the first six periods from all patients who completed at least six but no more than eight periods. Thus, in either block, all the retained data included exactly twice as many treatment observations as control observations for each patient.

As in the preceding example, their test statistic was the sum of the observations labeled as controls. But the relabeling was conducted separately and independently within each block. This division into blocks ensured that the observations are exchangeable under the null hypothesis so that the resulting significance level is exact.

In the clinical study described above, 74 patients began the study, 9 patients, divided (2, 3, 4) among the three treatment sequences, completed at least the first 6 periods, and 61 completed all 9 periods. The contribution to the test statistic of those who completed all nine periods was 3.83*61*3 = 700.9. The contribution of those who completed at least six periods but no more than eight was 43.4. The value of our new test statistic for the data as labeled originally is 744. The values of the test statistic for 1,600 random relabelings of the data ranged from 1,255 to 1,490, with the 2.5th percentile equal to 1,487. Good and Xie (1988) concluded that there was a statistically significant difference between the two treatments.

Which Sets of Labels Should We Rearrange?

As we saw in the preceding sections, the sets of labels to be rearranged will depend upon the hypothesis to be tested. Microarrays record the amounts of dyes taken up by individual genes in subjects who display certain characteristics and in those who do not. The objective is to determine which of the hundreds of thousands of genes are responsible for the characteristic. Once a set of candidate genes is in hand, the null hypothesis can be formulated in two different ways:

1. The genes in a selected gene set show the same pattern of associations with the characteristic of interest as the rest of the genes do.
2. The selected gene set does not contain any genes whose expression levels are associated with the characteristic of interest.

To test these hypotheses, the data are first put in a matrix whose rows are genes and whose columns are samples. Next, a measure of association ti between each gene i and the phenotype of interest is calculated. The test statistic T is formed by summing over all the measures of association ti for the genes in the set of interest.

To test the first hypothesis, Tian et al. (2005) propose that we form the permutation distribution by rearranging the rows of the data matrix and computing T for each rearrangement. To test the second hypothesis, we rearrange the columns of the matrix each time.

Determining Sample Size

As the sensitivity of our tests and the precision of our estimates are an increasing function of sample size, we always should take as large a sample as we can afford, providing the gain in power from each new observation is worth the expense of gathering it.

If we aren't sure about the underlying distribution, we draw a bootstrap sample with replacement from the data we already have in hand. Two cases arise:

1. We have preliminary samples from both the treated and untreated groups.
2. We have a preliminary sample from the untreated group only.

In the latter case, we create an artificial sample by adding (or subtracting) the smallest effect of practical interest from each of the observations in the original sample.

We compute the test statistic for the bootstrap samples, and note whether we accept or reject. We repeat the entire process 50 to 400 times (50 times initially, when we are just trying to get a rough idea of the correct sample size, 400 times when we are closing in on the final value). The number of rejections divided by the number of simulations provides us with an estimate of the power for the specific experimental design and our initial sample sizes. If the power is still too low, we increase the bootstrap sample sizes and repeat the preceding simulation process.

Missing Combinations

Makinodan et al. (1976) studied the interaction between spleen cells derived from old and young mice. In one set of trials, they studied the immune response of cells derived entirely from the spleens of young mice; in a second, the cells came from the spleens of old mice; and in a third, they came from mixtures of the two. Untested, for it seemed to these investigators unnecessary, was the control response in the absence of spleen cells. As a result, their design was unbalanced.

		Old Cells	
		Yes	No
Young Cells	Yes	Data	Data
	No	Data	

Let $X_{i,j,k}$ denote the response of the kth sample taken from a population of type i,j ($i = 1 = j$: controls; $i = 2, j = 1$: cells from young animals only; $i = 1, j = 2$: cells from old animals only; $i = 2 = j$: mixture of cells from old and young animals). For lymphocytes taken from the spleens of young animals,

$$X_{2,1,k} = \mu + \alpha + e_{2,1,k}$$

for the spleens of old animals,

$$X_{1,2,}k = \mu - \alpha + e_{1,2,k}$$

and for a mixture of p spleens from young animals and $(1 - p)$ spleens from old animals, where $0 \le p \le 1$,

$$X_{2,2,k} = p(\mu + \alpha) + (1 - p)(\mu - \alpha) - \gamma + e_{2,2,k}$$

where the $e_{2,2,k}$ are independent values.

Makinodan et al.'s (1976) primary interest was the possible interaction between the cells from the different age groups, for he knew beforehand that there would be differences in the results for old and young animals, that is, $\alpha > 0$. If the interaction term γ were 0, then one could infer the two cell populations did not interact. $\gamma < 0$ meant there were excess lymphocytes in young populations, while $\gamma > 0$ suggests the presence of suppressor cells in the spleens of older animals.

If the design were balanced, or if one could be sure that the effect μ in the absence of lymphocytes was 0, then the statistic of choice would be

$$S =| \bar{X}_{22.} - p\bar{X}_{21.} - (1-p)\bar{X}_{12.} |$$

But the design is not balanced, with the result that the main effects in which we are not interested are confounded with the interaction with which we are.

The bootstrap provides a solution: Draw an observation at random and with replacement from the set $\{x_{10}k\}$; label it x_{10}^*. Similarly, draw the bootstrap observations x_{01}^* and x_{11}^* from the sets $\{x_{01k}\}$ and $\{x_{11k}\}$. And let

$$\gamma^* = p\bar{X}_{2,1}^* + (1-p)\bar{X}_{1,2}^* - \bar{X}_{2,2}^*$$

Repeat this resampling procedure several hundred times, obtaining a bootstrap estimate γ^* of the interaction each time you resample. Use the resultant set of bootstrap estimates to obtain a confidence interval for γ. If 0 belongs to this confidence interval, accept the hypothesis of additivity; otherwise, reject.

As Makinodan et al. (1976) conducted many replications of this experiment for varying values of p with comparable results, they could feel confident in concluding that young spleens have an excess of lymphocytes.

Summary

In this chapter, you learned the principles of experimental design: to block or measure all factors under your control, and to randomize with regard to factors that are not. You learned to analyze balanced k-way designs and balanced r-by-c designs for main effects. You learned to use the Latin square to reduce sample size, to apply permutation methods to crossover designs to correct for carryover effects, and to use bootstrap methods when designs are not balanced.

To Learn More

For more on the principles of experimental design, see Fisher (1935), Kempthorne (1955), Wilk and Kempthorne (1956, 1957), Scheffe (1959), and Maxwell and Cole (1991). Further sample-size guidelines are provided in Shuster (1993).

Permutation tests have been applied to a wide variety of experimental designs, including clinical trials (Lachin, 1988a, 1988b), covariance (Peritz, 1982), crossovers (Shen and Quade, 1986; Good and Xie, 2008), factorial designs (Loughin and Noble, 1997; see Chapter 8), growth curves (Foutz et al., 1985; Zerbe, 1979a, 1979b), matched pairs (Peritz, 1985; Welch, 1987; Welch and Guitierrez, 1988; Rosenbaum, 1988; Good, 1991), randomized blocks (Wilk, 1955; Tang et al., 2009), restricted randomization (Smythe, 1988), and sequential clinical trials (Wei, 1988; Wei et al., 1986). Optimal permutation tests for analyzing group randomized trials, in which the outcomes within each group are often correlated, are developed by Braun and Feng (2001).

Mapleson (1986) applied the bootstrap to the analysis of clinical trials. See also Romano (1988).

Exercises

1. Design an experiment.
   a. List all the factors that might influence the outcome of your experiment.
   b. Write a model in terms of these factors.
   c. Which factors are under your control?
   d. Which of these factors will you use to restrict the scope of the experiment?
   e. Which of these factors will you use to block?
   f. Which of the remaining factors will you neglect initially, that is, lump into your error term?
   g. How will you deal with each of the remaining covariates?
   h. How many subjects/items will you observe in each subcategory?
   i. Write out two of the possible assignments of subjects to treatment.
   j. How many possible assignments are there in all?

2. A known standard was sent to six laboratories for testing.
   a. Are the results comparable among laboratories?

Date	Laboratory					
	A	B	C	D	E	F
1/1	221.1	208.8	211.1	208.3	221.1	224.2
1/2	224.2	206.9	198.4	214.1	208.8	206.9
1/3	217.8	205.9	213.0	209.1	211.1	198.4

   b. The standard was submitted to the same laboratories the following month. Are the results comparable from month to month?

	Laboratory					
Date	A	B	C	D	E	F
2/1	208.8	211.4	208.9	207.7	208.3	214.1
2/2	212.6	205.8	206.0	216.2	208.8	212.6
2/3	213.3	202.5	209.8	203.7	211.4	205.8

3. Potted tomato plants in groups of six were maintained in one of two levels of artificial light and two levels of water. What effects, if any, did the different levels have on yield?

Light	Water	Yield	Light	Water	Yield
1	1	12	2	1	16
1	1	8	2	1	12
1	1	8	2	1	13
1	2	13	2	2	19
1	2	15	2	2	16
1	2	16	2	2	17

4. a. Are the two methods of randomization used in the tomato trials equivalent?

   b. Suppose you had a six-sided die and three coins. How would you assign plots to one of four treatments? Three rows and two columns? Eight treatments?

5. Four containers, each with 10 oysters, were randomly assigned to four stations, each kept at a different temperature in the wastewater canal of a power plant. The containers were weighed before treatment and after 1 month in the water. Were there statistically significant differences among the stations? Can these data be analyzed by k-sample comparison methods?

Treatment	Initial	Final	Trt	Initial	Final
1	27.2	32.6	3	28.9	33.8
1	31.0	35.6	3	23.1	29.2
1	32.0	35.6	3	24.4	27.6
1	27.8	30.8	3	25.0	30.8
2	29.5	33.7	4	29.3	34.8
2	27.8	31.3	4	30.2	36.5
2	26.3	30.4	4	25.5	30.8
2	27.0	31.0	4	22.7	25.9

6. You can increase the power of a statistical test in three ways: (a) making additional observations, (b) making more precise observations, and (c) adding covariates. Discuss this remark in the light of your own experimental efforts.

7. A pregnant animal was accidentally exposed to a high dose of radioactivity. Her seven surviving offspring (one died at birth) were examined for defects. Three tissue samples were taken from each animal and examined by a pathologist. What is the sample size?

8. Should your tax money be used to fund public television? When a random sample of adults were asked for their views on a 9-point scale (1 is very favorable and 9 is totally opposed), the results were as follows:

   3, 4, 6, 2, 1, 1, 5, 7, 4, 3, 8, 7, 6, 9, 5

   The first 10 of these responses came from Manhattan, and the last 5 came from Yonkers. Are there significant differences between the two areas? Given that two-thirds of the population live in New York City, provide a point estimate and confidence interval for the mean response of the entire population. Hint: Modify the test statistic so as to weight each block proportionately.

9. Using the insecticide data in Table 6.8, test for independence of action using only the zero and highest dose level of the first drug. Can you devise a single test that would utilize the data from all dose levels simultaneously?

10. In how many different ways can we assign nine subjects to three treatments, given equal numbers in each sample? What if we began with six men and three women and wanted to block our experiment?

11. Take a second look at the data of Makinodan et al.'s immunology experiment. We seem to have three different samples with three different sample sizes, each drawn from a continuous domain of possible values. Or do we? From the viewpoint of the bootstrap, each sample represents our best guesstimate of the composition of the larger population from which it was drawn. Each hypothetical population appears to consist of only a finite number of distinct values, a different number for each of the different populations. Discuss this seeming paradox.

12. Without thinking through the implications, you analyze your data from a matched pairs experiment as if you had two independent samples and obtain a significant result. Consequently, you decide not to waste time analyzing the data correctly. Right or wrong? (Hint: Were the results within each matched pair correlated? What if one but not both of the observations in a matched pair were missing?)

13. **Review question** (Chapters 1–6). Chernick (2008, pp. 67–69) describes the use of a bootstrap to test whether a new model of a steroid-eluting pacemaker lead would result in at least a 0.5 V reduction in the mean capture threshold. Previous trials with the existing model of

lead had shown that the voltages were not normally distributed. Nonetheless, the data were assumed to be normal for this purpose.

Unequal sample sizes were employed, with 99 patients in the experimental group and 33 in the control group. Not surprisingly, patients were lost to follow-up during the study, and observations were made on only 89 patients in the experimental group and 29 in the control group. Five thousand bootstrap samples were drawn to obtain a confidence interval for the difference in voltage.

A Wilcoxon rank was used as an alternative to the bootstrap.

Find at least four things wrong with the preceding design and analysis.

7

Categorical Data

In many experiments and in almost all surveys, the results fall into categories: male vs. female, in favor vs. against vs. undecided. The corresponding hypotheses concern proportions: "blacks are as likely to be Democrats as they are to be Republicans." Or, "we may combine the success rates from several different clinical sites." Sometimes the categories can be ordered as when survey responses are recorded on a Likert scale (*hate, dislike, indifferent, like, love*). In this chapter, you learn to test hypotheses concerning both unordered and ordered categorical data.

Fisher's Exact Test

As an example of a contingency table, suppose the results of a microarray analysis were summarized as shown in Table 7.1.

The 9 denotes the nine differentially expressed genes that were in the gene set of interest, the 1 the remaining differentially expressed gene, and so forth. The four marginal totals, or *marginals*, are 10, 14, 13, and 11. The total number of differentially expressed genes in the study is 10; the total number of genes that were not is 14, and so forth.

We see in this table an apparent difference in the expression rates of genes that were in and not in the data set of interest: 9 in 10 vs. 4 in 14. Is this difference statistically significant?

The answer is yes. Let's see why, paralleling the reasons advanced by Fisher (1935). Table 7.1 has several fixed elements; its marginal totals are:

- The total number of genes that were differentially expressed, 10
- The total number of genes that were not, 14
- The total number in the gene set, 13
- The total number not in the gene set, 11

These totals are immutable; no swapping of labels will alter the total number of individual genes or the numbers that were differentially expressed. But these totals do not determine the contents of the table, as can we seen in Tables 7.2 and 7.3, which have identical marginal totals.

TABLE 7.1

	In Gene Set	Not in Gene Set	Marginal Total
Differentially expressed	9	1	10
Not differentially expressed	4	10	14
Marginal total	13	11	24

TABLE 7.2

	In Gene Set	**Not in Gene Set**	**Marginal Total**
Differentially expressed	10	0	10
Not differentially expressed	3	11	14
Marginal total	13	11	24

TABLE 7.3

	In Gene Set	**Not in Gene Set**	**Marginal Total**
Differentially expressed	8	2	10
Not differentially expressed	5	9	14
Marginal total	13	11	24

Table 7.2 makes a strong case for the diagnostic value of the selected gene set, even stronger than our original observations. In Table 7.3, whether or not a gene is in the designated set seems less important than in our original table.

These tables are not equally likely, not even under the null hypothesis. Table 7.2 could have arisen in any of 13 choose 3 ways, and Table 7.3 in any of $\binom{13}{8}\binom{11}{2}$ ways.

If the survival rates were the same for both sexes, then each of the redistributions of labels to subjects, that is, each of the N possible contingency tables with these same four fixed marginals, is equally likely, where

$$N = \binom{24}{10}$$

How did we get this value for N? The component terms are taken from the hypergeometric distribution:

$$\sum_{x=0}^{t} \binom{m}{x}\binom{n}{t-x} \Big/ \binom{m+n}{t} \tag{7.1}$$

TABLE 7.4

	Category 1	Category 2	Total
Category A	x	$M - x$	m
Category B	$t - x$		n
Total	t		$M + n$

where n, m, t, and x occur as the indicated elements in the 2×2 contingency in Table 7.4.

If the selected set of genes can not be distinguished from the remaining genes, then all tables with the marginals m, n, and t are equally likely, and

$$\sum_{k=0}^{t-x} \binom{m}{t-k}\binom{n}{k}$$

are as or more extreme. We count the more extreme tables when determining a *p*-value, because we know that if we are going to reject a hypothesis based on the table at hand, we also would have rejected the hypothesis had we observed a table that was more extreme.

In our example, $m = 13$, $n = 11$, $x = 9$, and $t = 10$, so that

$$\binom{14}{10}\binom{10}{1}$$

of the N tables are as extreme as our original table and

$$\binom{14}{11}\binom{10}{0}$$

are more extreme. The resulting sum is still only a very small fraction of the total N, so we conclude that a difference in differential expression as extreme as the difference we observed in our original table is very unlikely to have occurred by chance. We reject the null hypothesis and accept the alternative that the selected set of genes has greater diagnostic value.

Computing Fisher's Exact Test

It would be difficult to find statistical software that does *not* include Fisher's exact test. Thus, only the R code is provided in this section. To execute most of the other procedures described in this chapter, it is necessary to download a trial copy of StatXact™ from http://www.cytel.com/Downloads/Default.asp.

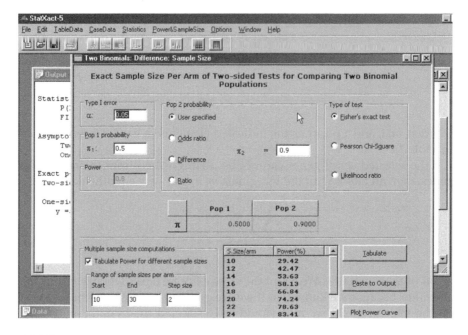

FIGURE 7.1
Using StatXact to determine required sample size.

R

To simplify the programming, assume that the smallest marginal is in the first row and the smallest cell frequency is located in the first column, and that we have the actual cell frequencies.

```
data =c(f11, f12, f21, f22)
m = data[2] + data[4]
n = data [1] + data [3]
t = data[1] + data[2]
ntab=0
for (k in 0:data[1]) ntab = ntab + comb(m,t-k)*comb(n,k)
ntab/comb(m+n,t) #prints the p-value for Fisher's Exact Test
```

where fact = function(n)prod (1:n) and comb = function (n,t) fact(n)/(fact(t)*fact(n-t)).

In S-PLUS, we would substitute choose() for comb().

Two-Tailed Tests

In the preceding example, we rejected the null hypothesis because only a small fraction of the possible tables were as extreme as the one we observed initially. This is an example of a one-tailed test. Or is it? Wouldn't we have

TABLE 7.5

	In Gene Set	Not in Gene Set	Marginal Total
Differentially expressed	0	10	10
Not differentially expressed	13	1	14
Marginal total	13	11	24

been just as likely to reject the null hypothesis if we had observed a table similar to Table 7.5?

In fact, we would probably have discarded the information from the selected set of genes and taken a more searching look at the genes we'd overlooked. In determining the significance level when we have a two-sided alternative (the selected set of genes is of greater/lesser diagnostic value than the remaining genes), we must add together the total number of tables that lie in either of the two extremes or tails of the permutation distribution.

McKinney et al. (1989) reviewed some 70 plus articles that appeared in six medical journals. In over half these articles, Fisher's exact test was applied improperly. Either a one-tailed test had been used when a two-tailed test was called for, or the authors of the paper simply hadn't bothered to state which test they had used.

When you design an experiment, decide at the same time whether you wish to test your hypothesis against a two-sided or a one-sided alternative. A two-sided alternative dictates a two-tailed test; a one-sided alternative dictates a one-tailed test.

As an example, suppose we decide to do a follow-on study with additional subjects to confirm our original finding that the selected gene set is of greater diagnostic value. In this follow-on study, we have a one-sided alternative. Thus, we would analyze the results using a one-tailed test rather than the two-tailed test we applied in the original study.

Unfortunately, it is not obvious which tables should be included in the second tail. Is Table 7.5 as extreme as Table 7.2? We need to define a test statistic to serve as a basis of comparison. One commonly used measure is the Pearson χ^2 statistic defined for the 2 × 2 contingency table after eliminating elements that are invariant under permutations as $[x - tm/(m + n)]^2$. This statistic is proportional to the square of the difference between what one would expect to observe if the survival rates are the same, that is, $tm/(m + n)$ and the frequency x actually observed. For Table 7.1, this statistic is approximately 13; for Table 7.5, it is approximately 29. As an exercise, use the chi-square criteria to show that Table 7.6 is more extreme than Table 7.1, but Table 7.7 is not.

Pearson's χ^2 is far from being our only choice as a test statistic; among the other choices are the likelihood ratio

$$x\log[xtm/(m + n)]$$

and Fisher's statistic

TABLE 7.6

	In Gene Set	Not in Gene Set	Marginal Total
Differentially expressed	1	9	10
Not differentially expressed	12	2	14
Marginal total	13	11	24

TABLE 7.7

	In Gene Set	Not in Gene Set	Marginal Total
Differentially expressed	2	8	10
Not differentially expressed	11	13	14
Marginal total	13	11	24

TABLE 7.8

	In Gene Set	Not in Gene Set	Marginal Total
Differentially expressed	7	3	10
Not differentially expressed	6	8	14
Marginal total	13	11	24

$$-2\log(h[y]) - \log[2.51(m + n)^{-3/2}(mnt)^{1/2}(m + n - t)^{1/2}]$$

where $h[y]$ is the proportion of all tables with the same marginals that have precisely the same four entries. For very large samples with a large number of observations in each cell, all three statistics lead to the same conclusion.

Borderline Significance

A problem with any of the methods we've used so far is that we're very unlikely to observe 0.05 or 0.01 exactly. If the number of observations is small, p-values may jump from 0.040 (as in Table 7.3) to 0.185 (as in Table 7.8) as a result of a single additional case.

What is the appropriate criteria for rejection of the null hypothesis? Limiting ourselves to 4% of the tables means we may accept when we should reject. Rejecting for 18% means we are rejecting far more often than we should. Here are six alternatives:

1. Deliberately err on the conservative side, that is, accept the null hypothesis absent strong evidence to the contrary. See Boschloo (1970) and McDonald et al. (1977) for some slight improvements on this approach.

2. Randomize on the boundary (Lehmann, 1986). If you get a *p*-value of 0.185 and the next closest value would have been 0.040, let the computer choose a random number between 0 and 1 for you. If this number is less than (0.05 − 0.04)/(0.185 − 0.040), reject the hypothesis at the 5% level; accept it otherwise. This is a great technique for abstract mathematics, but I don't recommend it for use in practice.

3. Use the mid-*p*-value. Let *p* be equal to half the probability of the table you actually observe plus all of the probability of more extreme results.

4. Make use of a backup statistic; see Streitberg and Roehmel (1990) and Cohen and Sackrowitz (2003).

5. Conduct a sensitivity analysis (Dupont, 1986). Add a single additional case to one of the cells (say the cell that has the most observations already, so your addition will have the least percentage impact). Does the *p*-value change appreciably? This approach is particularly compelling if you are presenting statistical evidence in a courtroom, as it turns impersonal percentages into individuals; see Good (2001).

6. Let your audience decide. Present them with the data and the *p*-value you calculated. Let them make up their own minds as to whether it is a significant result or not. See Kempthorne (1977).

Is the Sample Large Enough?

If we did *not* reject the null hypothesis, is this because the sample wasn't large enough to detect an effect? Of course, we ought to have addressed this issue *before* we collected the data.

Consider Table 7.9 in which we've recorded the results of a comparison of two drugs. It seems obvious that drug B offers significant advantages over drug A. If we look up the observed value of the chi-squared statistic in a table of the chi-square distribution, we'll get an erroneous *p*-value of 3%. Fisher's exact test yields the correct one-sided *p*-value, 7%.

Actually, we were quite fortunate in getting a *p*-value this small. Suppose the probability of a response with drug A really is 50% and the probability of a response with drug B is 0.9, using the Stata command, sampsi 0.5 0.9, *n1*(10) *r*(1), we learn that the power of Fisher's exact test to detect this alternative with this small a sample size is only 29%!

TABLE 7.9

	Drug A	Drug B
Response	5	9
No response	5	1

**USING STATA TO ESTIMATE THE POWER
OF FISHER'S EXACT TEST**

```
. sampsi 0.5 0.9, n1(10) r(1)
Estimated power for two-sample comparison of proportions
Test Ho: p1 = p2, where p1 is the proportion in
population 1
and p2 is the proportion in population 2
Assumptions:
alpha = 0.0500 (two-sided)
p1 = 0.5000
p2 = 0.9000
sample size n1 = 10
n2 = 10
n2/n1 = 1.00
Estimated power:
power = 0.2907
```

How large a sample size would we need to obtain a power of 80% for comparing proportions of 0.5 and 0.9? According to StatXact, we'd need at least 24 observations in a balanced design.

The p-value obtained from the chi-square distribution is often quite different from that provided by Fisher's exact test. As its name suggests, Fisher's exact test provides exact significance levels even for small samples, while the chi-square distribution represents an approximation that provides exact significance levels only for a quite large one. In this context, "small" and "large" are determined by the individual table entries and not by the overall sample size.

Odds Ratio

The *odds ratio* may be used to make a more powerful statement as to our findings, such as "men are twice as likely as women to get a good-paying job," or "women under 30 are twice as likely as men over 40 to receive an academic appointment."

For the data of Table 7.10, the odds ratio (1/9)/(10/4) is 0.044 with a 90% confidence interval using the method of Gart (1970) extending from 0.002 to 0.44.

In the discrimination case of *Fisher v. Transco Services of Milwaukee* (1992), the plaintiffs claimed that Transco was 10 times as likely to fire older employees. Can we support this claim with statistics? The Transco data are in Table 7.10.

TABLE 7.10

	Transco Employment	
Outcome	Young	Old
Fired	13	1
Retained	13	11

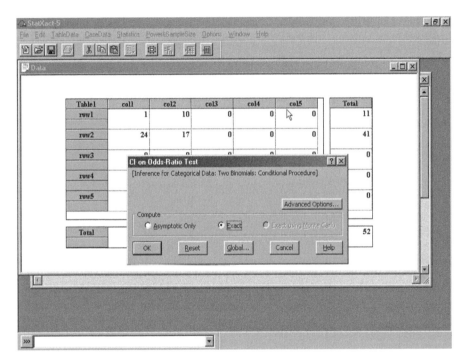

FIGURE 7.2
Using StatXact to determine a confidence interval for an odds ratio.

Let π_y denote the probability of firing a young person and π_o the probability of firing an older person. We want to go beyond testing the null hypothesis $\pi_y = \pi_o$ to determine a confidence interval for the odds ratio

$$\frac{\pi_o}{1-\pi_o} \Big/ \frac{\pi_y}{1-\pi_y}$$

Figure 7.2 illustrates the use of StatXact, a statistical package whose emphasis is the analysis of categorical and ordinal data, to determine a confidence interval for the odds ratio.

We choose in turn "Statistics," "Two Binomials," and "CI. Odds Ratio" from successive StatXact menus. Based on the results depicted in the accompanying box, we can tell the judge that older workers were fired at a rate at least 1.65 times that at which younger workers were discharged (and perhaps at a far greater rate).

ODDS RATIO OF TWO BINOMIAL PROPORTIONS

StatXact Output
Datafile: C:\EXAMPLES\TRNSCO.CY3
Statistic based on the observed 2 by 2 table:

```
Binomial proportion for column <young > : pi_1 = 0.04000
Binomial proportion for column <Old > : pi_2 = 0.3704
( pi_2 )/( 1-pi_2 )
Odds Ratio =————————-= 14.12
( pi_1 )/( 1-pi_1 )
Results:
Method P-value(2-sided) 95.00% Confidence Interval
Asymp (Mantel-Haenszel) 0.0157 ( 1.649, 120.9)
Exact 0.007145 ( 1.649, 637.5)
```

Stratified 2 × 2s

In expanding the range of applications of permutation tests to multiple contingency tables as well as tables with more than two rows and two columns (and, as we shall see in a later section, with more than two factors) we encounter two sets of problems:

1. The large number of calculations that are required
2. The many possible choices for a test statistic

The first of these problems was resolved through the efforts of Cyrus Mehta and his colleagues, who applied a combination of network analysis and dynamic programming to reduce the number of calculations that were required and to increase the efficiency of the calculations. An example of one such application is the test developed by Mehta et al. (1988) for the homogeneity of odds ratios in stratified 2 × 2 tables.

In trying to develop a cure for a relatively rare disease, we face the problem of having to gather data from a multitude of clinical sites, each with its own set of procedures and its own way of executing them. Before we can combine the data, we must be sure the odds ratios across the sites are approximately the same. Similar problems are encountered in studies where test subjects may use one of several different test apparatuses or be examined by one of several different examiners. Until we've performed an initial test of equivalence, results from different machines or the different technicians who operated them cannot be combined.

To obtain a Monte Carlo estimate of the significance level for Mehta et al.'s (1988) test of equivalence, download a copy of StatXact from http://cytel.com. Pull down StatXact's menus and select in turn "Statistics," "Stratified 2x2 Tables," and "Homogeneity of Odds Ratios" tables.

Controlling the False Discovery Rate

When applying categorical methods to high-throughput biological screens or to subsets of multidimensional contingency tables, both of which entail testing multiple hypotheses, we are guaranteed to encounter multiple false positives. Computer code for controlling the false discovery rate for multiple Fisher's exact tests is provided by Carlson et al. (2009).

Here is their algorithm for computing Fisher scores for all permutations of a table in pseudocode form:

```
AllFisherScores = function (t){
#Input: Contingency table t = (a; b; c; d)
#Output: Mapping of Fisher scores to hypergeometric
probabilities
D = min(a; d)
DV= c(-D,D,D,-D)
t₀ =t-DV
Vd=c(1,-1,-1,0)
L = p₀ = c(pr(T = t₀ | H = 0)) #requires additional function
for (i in 0 : (min(a; d) + min(c + d)){
tᵢ₊₁ = tᵢ + Vd
#compute pr(tᵢ₊₁ | H = 0)
pᵢ₊₁ = pᵢ*(bᵢ-cᵢ)/[(aᵢ+1)*(dᵢ+1)]
L=c(L,pi+1)
}
sort (L)
FisherScore = 0
for ( j in 0 : ){
FisherScore = FisherScore + L[i]
M[FisherScore] = L[i]
return M
```

Unordered $r \times c$ Contingency Tables

Halter (1969) showed we can find the probabilities of any individual $r \times c$ contingency table through a straightforward generalization of the hypergeometric distribution given in Equation 7.1. An $r \times c$ contingency table consists of a set of frequencies

$$\{f_{ij}, 1 \le i \le r; 1 \le j \le c\}$$

with row marginals $\{f_{i.}, 1 \le i \le r\}$ and column marginals $\{f_{.j}, 1 \le j \le c\}$.

Suppose once again we have mixed up the labels. To make matters worse, this time every item/subject is to be assigned both a row label ($f_{1.}$ of which are labeled row 1, $f_{2.}$ of which are labeled row 2, and so forth) and a column label. Let P denote the probability with which a specific table assembled at random will have these exact frequencies.

$P = Q/R$ where

$$Q = \prod_{i=1}^{r} f_{i.}! \prod_{j=1}^{c} f_{.j}! f_{..}! \text{ and } R = \prod_{i=1}^{r} \prod_{j=1}^{c} f_{ij}!$$

The Freeman and Halton (1951) test is an obvious extension of Fisher's exact test based on the proportion p of tables for which P is greater than or equal to P_o for the original table.

What is not so obvious is whether this extension offers any protection against the alternatives of interest. Just because one table is less likely than another under the null hypothesis does not mean it is going to be more likely under the alternatives of interest to us. Consider the 1×3 contingency table

$$f_1 \quad f_2 \quad f_3$$

which corresponds to the multinomial with probabilities $p_1 + p_2 + p_3 = 1$; the table whose entries are $f_1 = 1, f_2 = 2, f_3 = 3$ argues more in favor of the null hypothesis $p_1 = p_2 = p_3$ than the ordered alternative $p_1 > p_2 > p_3$.

The classic statistic for independence in a contingency table with r rows and c columns is the Pearson chi-square statistic:

$$\chi^2 = \sum_{i=1}^{r} \sum_{j=1}^{c} (f_{ij} - E[f_{ij}])^2 / E[f_{ij}],$$

where $E(f_{ij})$ is the number of observations in the ijth category one would expect on theoretical grounds.

Cressie and Read (1984) note that the Pearson chi-square statistic is one of the many power divergence statistics $CR(\lambda)$ that may be used to ascertain goodness of fit:

$$CR(\lambda) = \sum_{i=1}^{m^k} \frac{1}{\lambda(\lambda+1)} O(x_i)[O(x_i)^{\lambda} / E(x_i) - 1]$$

where $O(x_i)$ is the observed frequency of the response pattern x_i and $E(x_i)$ is the expected frequency.

With very large samples, this statistic has the chi-square distribution with $(r-1)(c-1)$ degrees of freedom. But in most practical applications, the chi-square distribution is only an approximation to the distribution of this statistic and is notoriously inexact for small and unevenly distributed samples.

We recommend that you make use of the permutation statistic based on the proportion of tables for which the Pearson chi-square statistic is greater than or equal to its value for the original table. This resampling method provides an exact test while possessing all the advantages of the original chi-square. The distinction between the two approaches is that with the chi-square test as it is described in most textbooks, we look up the significance level in a table, while with the permutation statistic, we derive the significance level from the permutation distribution. With large samples, the two approaches are equivalent, as the permutation distribution converges to the tabulated distribution (see Bishop et al., 1975, chapter 14).

Alternative tests of association include the likelihood ratio test based on the statistic

$$p_L = 2 \sum_{i=1}^{r} \sum_{j=1}^{c} f_{ij} \log[(f_{ij}f_{..})/(f_{i.}f_{.j})]$$

and Cramer's *V*:

$$V = \sqrt{\chi^2 / f_{..}(\min[r,c]-1)}$$

Again, we recommend you make use of the permutation distributions of these statistics rather than rely on large sample approximations.

Test of Association

Regardless of which statistic we choose, these three permutation tests have one of the original chi-square test's disadvantages: while they offer global protection against a wide variety of alternatives, we can always find a more powerful test against any specific alternative. Row and column categories are treated equally, and no attempt is made to distinguish between cause and effect. To address this deficiency, Goodman and Kruskal (1954) introduce an asymmetric measure of association for nominal scale variables called tau, τ, which measures the proportional reduction in error obtained when one variable, the "cause" or independent variable, is used to predict the other, the "effect" or dependent variable.

If the different values of the independent variable correspond to different rows, let

$$\tau = \frac{\sum_j \max_i f_{ij} - \max_i f_{i.}}{\max_{ij} f_{..} - \max_i f_{i.}}.$$

TABLE 7.11

3	6	9
6	12	18

$\tau = 0$

TABLE 7.12

18	0	0
0	36	0

$\tau = 1$

TABLE 7.13

3	6	9
12	18	6

$\tau = 0.057$

$\tau = 0$ when the variables are independent; $\tau = 1$ when for each category of the independent variables, all observations fall into exactly one category of the dependent. These points are illustrated in the 2 × 3 examples in Tables 7.11 to 7.13.

To obtain this latter result from StatXact, select first "Statistics" and then "Nominal Response" (at the foot of the "Statistics" menu.). A permutation test of independence in this latter table is based upon the proportion of tables with the same marginals for which τ is less than 0.057.

Cochran's Q provides an alternate test for independence. Suppose we have I experimental subjects on each of whom we administer J tests. Let $y_{ij} = 1$ or 0 denote the outcome of the jth test on the ith patient; e.g., if the test is positive, y is set to 1 and is set to 0 otherwise. Define $R_i = jy_{ij}$; $C_j = iy_{ij}$.

$$Q = \frac{\sum_j (C_j - C.)^2}{R. - \sum_i R_i^2}$$

WHICH TEST SHOULD WE USE?

If there are exactly two rows and two columns, use Fisher's exact test.

If there are more than two rows and at least two columns and you want to test whether the relative frequencies are the same in each row and in each column, use the Freeman-Halton test use chi-square.

If there are more than two rows and at least two columns and you want to test whether the column frequencies depend on the row, use τ or Q.

Causation vs. Association

A lack of statistical significance may mean either that the sample size is too small or that the observed results may well have been the result of chance alone. On the other hand, a significant value for any of the above statistics only means that the row and column factors are associated. It does not mean a cause-and-effect relationship exists between them. They may both depend on a third variable omitted from the study.

Regrettably, the converse is also true. A third omitted variable may also result in two variables appearing to be dependent when the opposite is true; this is termed Simpson's paradox. Consider Table 7.14.

We don't need a computer program to tell us the treatment has no effect on the death rate. Or does it? Consider Table 7.15a and b, which result when we examine the men and women separately.

In the first of these tables, treatment reduces the male death rate from 0.43 to 0.38, and in the second, from 0.6 to 0.55. Both sexes show a reduction, yet the combined population does not!

The lesson to be learned is that one needs to think deeply about the underlying cause-and-effect relationships before analyzing data. In the preceding example, when we fail to account for the relative proportions of the sexes in the study population, we are led to a false conclusion.

TABLE 7.14

Sexes Combined

	Control	Treated
Alive	6	20
Dead	6	20

TABLE 7.15a

Men Only

	Control	Treated
Alive	4	8
Dead	3	5

TABLE 7.15b

Women Only

	Control	Treated
Alive	2	12
Dead	3	15

TABLE 7.16

Data Gathered by Graubard and Korn (1987)

	Maternal Alcohol Consumption (drinks/day)					
	0	<1	1 to 2	3–5	>6	Total
Malformation absent	17,066	14,464	788	126	37	
Malformation present	48	38	5	1	1	

Ordered Statistical Tables

In many surveys, respondents are asked to rate themselves on a discrete ordinal (that is, ordered) scale. As you can see from Table 7.16, which summarizes data gathered by Graubard and Korn (1987), in contrast to data that are measured on a continuous basis, ties with ordinal and categorical data are inevitable—the rule, not the exception. Thus, we need make use of a contingency table to analyze such results.

What value ought we to assign to each of the ordered categories? Are two drinks daily twice as bad as one drink? Among the leading choices for a scoring method are the following:

1. Category number: 1 for the first category, 2 for the second, and so forth
2. Mid-rank scores
3. Scores determined by the domain expert who collected the data—a biologist, a physician, a physiologist

To show how such scores might be computed, consider the following 1 × 2 contingency table:

	Alcohol Consumption	
Drinks/day	0	1 or 2
Frequency	3	5

The category scores are 1 and 2. The ranks of the eight observations are 1 through 3, and 4 through 8, so that the mid-rank score of those in the first category is 2, and in the second 7. User-chosen scores, based on alcohol consumption, might be 0 and 1.5.

To analyze grouped data such as that in Table 7.16 using R, we proceed as follows:

```
MalAbs=c(17066, 14464, 788, 126, 37)
MalPres=c(48, 38, 5, 1, 1)
NumCat=5
```

```
    Score = c(1: NumCat)
    Samp1=c(rep(Score[1], MalAbs[1]))
    for (i in 2: NumCat)Samp1 = c(Samp1, rep(Score[i],
MalAbs[i]))
    #Create Samp2 in similar fashion, then analyze as in Section
3.4.2.
```

```
#If you wish to use midrank scores instead, insert the
following code
    Score=numeric(NumCat)
    TotFreq=MalAbs +MalPres
    Total = 0
    for (i in 1: NumCat){
    Score[i]= Total +TotFreq[i]/2
    Total = Total +TotFreq[i]
    }
```

Partial Dependence

What if the table entries are interdependent? If π_{ij} is the expected frequency of individual observations in an $R \times C$ contingency table, then we may write

$$\pi_{ij} = \pi_{i.}\pi_{.j}(1 + \Sigma_m\lambda_m x_{im}y_{jm})$$

where $\pi_{i.}$ and $\pi_{.j}$ are the table marginals, $1 \le m \le M = \min(r, c) - 1$, and $\lambda_1 \ge \lambda_2 \ge \dots \ge \lambda_M$. In this canonical representation, the departure from independence is partitioned into M components, where the x's and y's are the scalings or score systems for the categories of the row and column variables, and λ is the canonical correlation between the row and column variables with the given scalings.

Consider the following hypotheses:

C0: $\lambda_1 = \lambda_2 = \dots = \lambda_M = 0$, that is, total independence

C1: $\lambda_2 = \dots = \lambda_M = 0$

C2: $\lambda_3 = \dots = \lambda_M = 0$

and so forth.

Reiczigel (1996) suggests the following testing procedure. First, model C0 is tested, and if it fits, the process ends. If it has to be rejected, testing continues with model C1 and so on, until one of the models turns out to be acceptable.

The main steps of testing model Ck are as follows:

1. Estimate the parameters (λ's, x's, and y's) by least squares.
2. Compute the entries in the contingency table, assuming model Ck is true.

3. Block bootstrap from this table and compute a test statistic t^* for each bootstrap sample.

4. Compare the actual value of t to a desired level critical value of the bootstrap distribution of the t^* values.

Correspondence Analysis

Correspondence analysis (CA) is a statistical visualization method for picturing the associations between the levels of a two-way contingency table. It is used to determine how similar the categories that compose the rows (or columns) are, as well as to determine which associations are strongest among the categories that define the rows and the categories that define the columns.

As in the preceding section, the first step in a CA is to obtain a canonical representation. At issue is whether slight perturbations in the initial contingency table will produce substantial alterations either in the extracted principal axes or in the configurations represented in the principal planes.

Two distinct types of bootstrap resampling, with quite different objectives and degrees of complexity, can be used for this assessment.

A total bootstrap carries out a complete CA for each bootstrap table built. Coordinates, eigenvalues, eigenvectors, absolute contributions, and relative contributions are obtained in each one of these analyses. The bootstrap table is built with the same sample size as the original table using the relative frequencies of the cells of the original table as estimates of the multinomial probabilities.

A partial bootstrap projects rows or columns of the bootstrapped tables on the subspace issued from CA of the original table, as illustrative or supplementary elements. This way, and contrary to the TB, the aim at the study of PB is only the analysis of the stability of the configurations and is less computationally intensive. See Álvarez et al. (2006) for an illustration of the method.

More than Two Rows and Two Columns

We need to consider two distinct cases: the first when the columns but not the rows of the table may be ordered (the other variable being purely categorical), and the second when both columns and rows can be ordered.

Singly Ordered Tables

Our approach parallels that of Chapter 6 where we describe a k-sample comparison of metric data. Our test statistic is

$$F_2 = \sum (S_i - \bar{S})^2$$

where

$$S_i = \sum g[j] f_{ij}$$

TABLE 7.17

Response to Chemotherapy

	None	Partial	Complete
CTX	2	0	0
CCNU	1	1	0
MTX	3	0	0
CTX + CCNU	2	2	0
CTX + CCNU + MTX	1	1	4

As in the case of the $2 \times c$ table, our problem is in deciding on the appropriate scoring function $g[]$.

Table 7.17 provides tumor regression data for five chemotherapy regimens. After questioning an oncologist (the domain expert), suppose we learned that a partial response corresponds to approximately 2 years in remission (about 100 weeks) and a complete response to an average of 3 years (150 weeks). As a result, we assign scores of 0, 100, and 150 to the ordered response categories.

To use StatXact to analyze the tumor data, we first click on "TableData" in the main menu, click on "Settings," then enter column scores 0, 100, and 150. To execute the analysis, we select in turn "Statistics," "Singly Ordered RxC Table," "ANOVA with Arbitrary Scores," and "Exact test method." The results are tabulated below.

**ANOVA TEST
(THAT THE 5 ROWS ARE IDENTICALLY DISTRIBUTED)**

Datafile: C:\EXAMPLES\TUMOR.CY3
Statistic based on the observed data:

```
The Observed Statistic = 7.507
Asymptotic p-value: (based on Chi-square distribution
with 4 df )
Pr { Statistic .GE. 7.507 } = 0.0747
Monte Carlo estimate of p-value :
Pr { Statistic .GE. 7.507 } = 0.0444
99.00% Confidence Interval = ( 0.0350, 0.0538)
```

Although our estimated significance level is less than 0.0444, a 99% confidence interval for this estimate, based on a sample of 3,200 possible rearrangements, does include values greater than 0.05. We can narrow this confidence interval by sampling additional rearrangements. This is because a Monte Carlo estimate is actually a binomial random variable $B(n, p)$, where n is the number of rearrangements and p is the true (but unknown) p-value. The variance of our estimate is $p(1 - p)/sqrt(n)$.

With a Monte Carlo of 10,000 sample tables, our estimate of the *p*-value is 0.0434, with a 99% confidence interval of (0.0382, 0.0486). Of course, the calculations take three times as long. When I first perform a permutation test, I use as few as 400 to 1,600 simulations. If the results are equivocal, as they are in this example, then and only then will I run 10,000 simulations.

Doubly Ordered Tables

In an $r \times c$ contingency table conditioned on fixed marginal totals, the outcome depends on the $(r - 1)(c - 1)$ odds ratios

$$\phi_{ij} = \frac{\pi_{ij}\pi_{i+1,j+1}}{\pi_{i,j+1}\pi_{i+1,j}}$$

where π_{ij} is the probability of an individual being classified in row *i* and column *j*.

In a 2×2 table, conditional probabilities depend on a single odds ratio, and hence one- and two-tailed tests of association are easily defined. An $r \times c$ table has a potential of $n(r - 1)(c - 1)$ sets of extreme values, two for each of the $(r - 1)(c - 1)$ odds ratios. An omnibus test for no association, e.g., χ^2, might have as many as 2^n tails.

Following Patefield (1982), we consider tests of the null hypothesis of no association between row and column categories $\phi_{ij} = 1$ for all *i, j* against the alternative of a positive trend $\phi_{ij} \geq 1$ for all *i, j*.

The principal test statistic considered by Patefield, also known as the linear-by-linear association test, is

$$\Lambda = \Sigma\Sigma f_{ij} r_i c_j$$

where $\{r_i\}$ and $\{c_j\}$ are user-chosen row and column scores. Whether it is statistically significant in any given case depends on the proportion of contingency tables with the same marginals in which its value is as or more extreme than its value for the original table.

Multidimensional Arrays

When survey results are tabulated, it is not unusual to have 10–20 response variables, each of which may take 2 to 5 values. It is also not unusual for the total number of observations N to be less than the total number of cells KM

in the contingency table, where K is the number of response variables, and M is the number of possible responses.

Two issues need to be resolved. What test statistic or statistics will be used? How will their distribution be determined?

Can the bootstrap provide the needed distribution? Bollen and Stine (1993) found that the percentile bootstrap failed when applied to the original data. Langeheine et al. (1996) successfully applied a parametric bootstrap. They treated the contents of each cell as if it were a multinomial variable whose parameters could be determined from the marginals. Their computations are greatly simplified by the use of a balanced bootstrap.

1. Put the data in matrix form with the rows of the table corresponding to subjects and the columns to variables. The entries in each cell of this table are the recorded values of the variables. Assume the table has the form $S \times K$, where S is the number of subjects.

2. Create a vector of length S containing the indices 1, 2, ..., S.

3. Build a vector **V** of length BS by replicating the previous vector B times, where B is the proposed number of bootstrap samples.

4. Randomly permute the elements of **V**.

5. Use the first S elements of the permuted vector to select subjects from the case table for the first bootstrap sample. Compute the test statistic for this sample.

6. Continue to select successive groups of S elements to form bootstrap samples and compute values of the test statistic until the vector **V** is exhausted.

Langeheine et al. (1996) made use of the log-likelihood ratio as their test statistic. The test statistics you employ will depend upon the hypotheses you wish to test.

Summary

In this chapter, you were introduced to the concept of a contingency table with fixed marginals, and shown you could test against a wide variety of general and specific alternatives by examining the resampling distribution of the appropriate test statistic. Among the test statistics you considered were Fisher's exact, Freeman-Halton, Pearson's chi-square, tau, Q, and linear-by-linear association. The latter two statistics are to be used when you can take advantage of an ordering among the categories.

To Learn More

Excellent introductions to the analysis of contingency tables may be found in Agresti (2002), and in the StatXact manual authored by Mehta and Patel (1983). Major advances in analysis by resampling means have come about through the efforts of Gail and Mantel (1977), Mehta and Patel (1983), Mehta et al. (1988), Baglivo et al. (1988), and Smith et al. (1996).

Santner and Snell (1980) derive confidence intervals for the difference and the ratios of the probabilities in a 2 × 2 table. To study a 2 × 2 table in the presence of a third covariate, see Bross (1964) and Mehta et al. (1985). For the application of Fisher's exact test to microarrays, see Goeman and Bühlmann (2007).

The power of the Freeman-Halton statistic in the $r \times 2$ case was studied by Krewski et al. (1984). Details of the calculation of the distribution of Cochran's Q under the assumption of independence are given in Patil (1975). For a description of some other, alternative statistics for use in $r \times c$ contingency tables, see Nguyen (1985). *Post hoc* analysis of a single cell of a contingency table via bootstrap means is described by Benjamini and Yekutieli (2001).

Pesarin and Salmaso (2010) review the combination of independent tests for univariate and multivariate ordered categorical data.

Exercises

1. Suppose we want to make use of the data relating tattoos and hepatitis C at http://statcrunch.com/5.0/index.php?dataid=339733.

    a. What hypothesis might we want to test?

    b. What assumptions about the data do we have to make in order to test this hypothesis?

    c. If these assumptions were satisfied, would you accept the null hypothesis?

2. In many epidemiological studies, each entry in the table is the result of an independent Poisson process and the total number of observations is itself a random variable. In a 2 × 2 table with categories ab, aB, Ab, and AB, one might want to test the hypothesis that the ratio $\lambda_{ab}/\lambda_{Ab} \leq \lambda_{aB}/\lambda_{AB}$. Would you use the same test as you would for comparing two binomials?

3. A recent report in the *New England Journal of Medicine* concerned a group of patients with a severe bacterial infection of their bloodstream who received a single intravenous dose of a genetically altered antibody.

a. In comparing death rates of the treated and untreated groups, should we use a one-tailed or a two-tailed test?

The report stated that those in the treated group had a 30% death rate compared with a 49% death rate for a group of untreated patients.

b. How large a sample size would you require using Fisher's exact test to show that such a percentage difference was statistically significant at the 5% level?

4. Will encouraging your child promote his or her intellectual development? A sample of 100 children and their mothers were observed and the children's IQs tested at 6 and 12 years. Before examining the data:

a. Do you plan to perform a one-tailed or two-tailed test?

b. What is the significance level of your proposed test?

	Mothers Encourage Schoolwork		
	Rarely	Sometimes	Never
IQ Increased	8	15	27
IQ Decreased	30	9	11

5. Does 1,2-dichloroethane induce tumors? Consider the following data evaluated by Gart (1986).

	Tumor	No Tumor
Treated	15	21
Control	2	35

6. Referring to the literature of your own discipline, see if you can find a case where an $r \times 2$ table with at least one entry smaller than 7 gave rise to a borderline p-value using the traditional chi-square approximation. Reanalyze this table using resampling methods. Did the authors use a one-tailed or a two-tailed test? Was their choice appropriate?

7. Show that once you have selected $(r - 1)(c - 1)$ of the entries in a contingency table with R rows and C columns the remainder of the entries are determined. (Hint: Solve in turn for 2×3, $2 \times c$, and $r \times c$ tables.)

8. Compared to men, do women tend to be Democrats?

	Party Identification in 1991		
	Democrat	Independent	Republican
Women	279	73	225
Men	165	47	191

9. Diseased trees were treated in a variety of ways. Do you feel the treatments were helpful? How many rows and columns in the following contingency table? Which, if any, of the factors is ordered?

	Control	Removal of Diseased Parts	Antibiotic
Died in one month	5	3	2
Survived 1 year	2	3	4
Survived 2–4 years	0	2	3

10. Based on the following table of results, would you conclude that treatment A is superior?

	A	B
Marked improvement	7	3
Moderate	15	9
Slight	16	14
No change	13	21
Worse	1	5

8

Multiple Hypotheses

One of the difficulties with large-scale studies like clinical trials as well as in the analysis of biomedical images, microarrays, and satellite imagery is that frequently so many variables are under investigation that one or more of them is practically guaranteed to be significant by chance alone. If we perform 20 tests at the 5% or 1/20 level, we expect at least one significant result on average. If the variables are related (and in most large-scale medical and sociological studies the variables have complex interdependencies), the number of falsely significant results could be many times greater.

In this chapter, you'll learn a variety of methods for controlling either the overall error rate or the false discovery rate.

Controlling the Family-Wise Error Rate

A resampling procedure outlined by Troendle (1995) allows us to work around the dependencies among tests. Suppose we have measured k variables on each subject, and are now confronted with k test statistics. To make these statistics comparable, we need to standardize them and render them dimensionless, dividing each by its respective L_1 *norm* or by its standard error. For example, if one variable, measured in centimeters, takes values like 144, 150, and 156, and the other, measured in meters, takes values like 1.44, 1.50, and 1.56, we might divide each of the first set of observations by 4, and each of the second set by 0.04.

Next, we order the standardized statistics by magnitude, that is, from smallest to largest. We also reorder and renumber the corresponding hypotheses. The probability that at least one of these statistics will be significant by chance alone at the 5% level is $1 - [1 - 0.05]^k$. But once we have rejected one hypothesis (assuming it was false), there will only be $k - 1$ true hypotheses to guard against rejecting.

1. Focusing initially on the largest of the k test statistics, repeatedly resample the data, with or without replacement, to determine the p-value.

2. If this p-value is less than the pr-determined significance level, then accept this hypothesis as well as all the remaining hypotheses.

3. Otherwise, reject the corresponding hypothesis, remove it from further consideration, and repeat steps 1 to 3.

Recent improvements to this method were documented by Somerville and Hemmelmann (2008).

Microarray Analysis

Microarray analysis presents a unique challenge in statistics as it is characterized by a small sample size n and a large number p of features (variables), often with $n \ll p$. When resampling methods are applied to microarray data, it is crucial to perform feature selection within each resampling step when estimating prediction errors—a process known as *honest* performance assessment (Dudoit and Fridlyand, 2003). With a huge number of features, the prediction rules contain two key steps: the feature selection and the class prediction step. Feature selection is administered prior to the class prediction step for every learning set. Failure to include feature selection in resampling steps results in serious downward bias in estimating prediction error and overly optimistic assessment of the prediction rule.

The Monte Carlo permutation approach proposed by McIntyre et al. (2000) lends itself to the analysis of data from microarrays when we want to detect changes in up to 100,000 single-nucleotide polymorphisms.

For a collection of genes, one set of gene variants is transmitted by the parents and a second set is not transmitted by the parents. Under the null hypothesis, the labels "transmitted" and "not transmitted" can be permuted for the sets of gene variants.

Let TDT denote the value of the test statistic for transmission at a single gene. Execute the following steps:

1. Calculate the maximum value of TDT; call this maximum TDT_{MAX} over all genes.
2. For each family, permute the "transmitted" and "not transmitted" labels randomly.
3. Calculate TDT_{MAX} for the permuted data.
4. If TDT_{MAX} for the permuted data is larger than the value of TDT_{MAX} from the original data, count 1; otherwise, count 0.
5. Repeat steps 2 to 4 M times.
6. Estimate the p-value as the total count from step 4 divided by the total number of shuffles M.

EEG Analysis

Analysis of the large, multidimensional data arrays produced by electroencephalographic (EEG) measurements of human brain function are similarly complicated due to the large number of variables and small number of subjects. Many time points are sampled in each experimental trial, so that making adjustment for multiple comparisons is mandatory. Given the typically large number of comparisons and the clear dependence structure among time points, simple Bonferroni type adjustments are far too conservative. A three-step approach has been proposed by Wheldon et al. (2007):

1. Summing univariate statistics across variables
2. Using permutation tests for treatment effects at each time point
3. Adjusting for multiple comparisons using permutation distributions to control family-wise error across the whole set of time points

See also Belmonte and Yurgelun-Todd (2001) for details of a similar analysis applied to magnetic resonance image analysis.

Controlling the False Discovery Rate

An alternate approach with large numbers of test is to control the false discovery rate (FDR), an approach first suggested by Benjamini and Hochberg (1995). When the tests are dependent, as is the case with microarrays, the subsampling approach proposed by Romano et al. (2008) is recommended.

In describing their procedure, we eliminate a large amount of technical detail. Let n be the size of the original sample and b the size of each of the subsamples, $b < n$. There is a total of $N = \binom{n}{b}$ possible subsamples. Let $T_{b,i,j}$ denote the value of the test statistic used to test the jth hypothesis, $j = 1, ..., J$, by making use of the ith subsample of size b. Let $T^*_{b,i,r}$ denote the rth largest of the test statistics $T_{b,i,1}, ..., T_{b,i,J}$.

After determining critical values $\{c_j\}$, a step-down procedure is employed: If $T^*_{b,i,J} < c_J$, then the procedure rejects no hypotheses; otherwise, the procedure rejects the hypothesis corresponding to $T_{b,i,J}$ and then "steps down" to the second most significant null hypothesis corresponding to $T_{b,i,J-1}$. If $T_{b,i,J-1} < c_{J-1}$, then the procedure rejects no further null hypotheses; otherwise, the procedure rejects the second most significant null hypothesis and then steps down to the third most significant hypothesis. The procedure continues in this fashion until either one rejects *all hypotheses* or one does not reject the hypothesis under consideration. More succinctly, a step-down multiple testing procedure rejects the j corresponding hypotheses where j is the largest integer that satisfies $T^*_{b,i,J} < c_{J-j}, ..., T^*_{b,i,j} < c_{J-j}$.

The critical values are determined recursively, beginning with c_1 such that

$$c_j = \inf_c \left\{ \frac{1}{N} \sum_{i=1}^{N} \sum_{k=1}^{j} \frac{k}{s-j+k} \; I\{T^*_{b,i,r} \geq c_r, T^*_{b,i,r-k+1} \geq c_{r-k+1}, T^*_{b,i,r-k} < c_{r-k}\} \leq \alpha \right\}$$

where $I\{\}$ is 1 if its argument is true and 0 otherwise.

I'd be happy to provide Dr. Wolf's C++ code for this procedure to anyone who is interested.

A straightforward adaptive step-down procedure due to Gavrilov et al. (2009) is claimed by these authors to be as powerful as the preceding algorithm while offering the advantage that it is easier to follow and computationally far simpler.

Suppose we wish to control the false discovery rate at $q < 1$. First, sort the p-values corresponding to the m univariate hypotheses as $p_{[1]} \leq \ldots \leq p_{[m]}$. Set the cutoff values at $c_i = iq/\{m + 1 - i[1 - q]\}$, $i = 1, \ldots, m$.

Let $k = \max \{1 \leq i \leq m$ with $p_{[j]} \leq c_j, j = 1, \ldots, i\}$. Reject the k hypotheses associated with $p_{[1]} \ldots p_{[k]}$ if k exists; otherwise, accept all hypotheses.

```
#rcode. p is a vector of length m containing the p values.
CFDR <- function(p, q) {
 m <- length(p);
 p <- sort(p);
 i <- 1;
 while(i<=m && p[i] <= i*q/(m+1-i*(1-q))) i <- i+1;
 i-1
}
```

Note that we've not specified how the p-values were determined or what tests were employed, as this will depend upon your specific application.

Software for Performing Multiple Simultaneous Tests

AFNI

AFNI incorporates permutation tests for use in analyzing neuroimages. Versions for Macs, PCs, and Unix computers may be downloaded from http://afni.nimh.nih.gov/afni/download/afni.

ExactFDR

The ExactFDR software package improves speed and accuracy of the permutation-based false discovery rate (FDR) estimation method by replacing the permutation-based estimation of the null distribution by generalization of the algorithm used for computing individual exact p-values. It provides a quick and accurate nonconservative estimator of the proportion of false positives in a given selection of markers. A Java 1.6 (1.5-compatible) version is available at http://sourceforge.net/projects/exactfdr.

NPCtest

To perform Pesarin's omnibus test, download a trial version from http://www.gest.unipd.it/~salmaso/NPC_TEST.htm

R

Permtest compares two groups of high-dimensional signal vectors derived from microarrays, for a difference in location or variability. Download from http://cran.r-project.org/web/packages/permtest/index.html.

Multtest provides nonparametric bootstrap and permutation resampling-based multiple testing procedures (including empirical Bayes methods) for controlling the family-wise error rate (FWER), generalized family-wise error rate (gFWER), tail probability of the proportion of false positives (TPPFP), and false discovery rate (FDR). Several choices of bootstrap-based null distribution are implemented (centered, centered and scaled, quantile transformed). Single-step and step-wise methods are available. Tests based on a variety of t- and F-statistics (including t-statistics based on regression parameters from linear and survival models, as well as those based on correlation parameters) are included. Results are reported in terms of adjusted p-values, confidence regions, and test statistic cutoffs. The procedures are directly applicable to identifying differentially expressed genes in DNA microarray experiments. To install this package for Windows or Macs, start R and enter:

```
source["http://bioconductor.org/biocLite.R"]
biocLite["multtest"]
```

SAS

A SAS macro to perform Pesarin's omnibus test may be downloaded from http://homes.stat.unipd.it/pesarin/NPC.SAS.

Testing for Trend

Suppose you conduct a small study to test the effect of a drug on 15 subjects. The subjects receive 0, 1, and 2 mg of the drug, and the presence or absence of 10 different side effects is noted for each subject. The data are taken from the SAS manual at http://support.sas.com/onlinedoc/913/docMainpage.jsp. Contrary to the statements made there, we shall suppose that the idea of testing for trend was conceived *prior* to our examining the data. Otherwise,

our stated significance level is a gross underestimate. For while SAS PROC
MULTTEST corrects for our conducting tests on the 10 side effects simultaneously, it does not correct for our having focused after the fact on the most
prominent of the many possible contrasts among the means of the three dose
groups.

```
data Drug;
input Dose$ SideEff1-SideEff10;
datalines;
0MG 0 0 1 0 0 1 0 0 0 0
0MG 0 0 0 0 0 0 0 0 0 1
0MG 0 0 0 0 0 0 0 0 1 0
0MG 0 0 0 0 0 0 0 0 0 0
0MG 0 1 0 0 0 0 0 0 0 0
1MG 1 0 0 1 0 1 0 0 1 0
1MG 0 0 0 1 1 0 0 1 0 1
1MG 0 1 0 0 0 0 1 0 0 0
1MG 0 0 1 0 0 0 0 0 0 1
1MG 1 0 1 0 0 0 0 1 0 0
2MG 0 1 1 1 0 1 1 1 0 1
2MG 1 1 1 1 1 1 0 1 1 0
2MG 1 0 0 1 0 1 1 0 1 0
2MG 0 1 1 1 0 1 1 1 1 1
2MG 1 0 1 0 1 1 1 0 0 1
;

proc multtest perm nsample=2000 seed=41287 notables pvals;
class Dose;
test ca[SideEff1-SideEff10/perm=2500];
contrast 'Trend' 0 1 2;
run;
```

THE SAS SYSTEM 12:29 THURSDAY, SEPTEMBER 30, 2004 1

```
The Multtest Procedure
Model Information
Test for discrete variables:            Cochran-Armitage
Exact permutation distribution used     Everywhere
Tails for discrete tests                Two-tailed
Strata weights                          None
P-value adjustment                      Permutation
Number of resamples                     2000
Seed                                    41287

Contrast Coefficients
Dose
Contrast 0MG 1MG 2MG
Trend 0 1 2
```

```
p-Values
Variable      Contrast    Raw        Permutation
SideEff1      Trend       0.1006     0.5205
SideEff2      Trend       0.3337     0.9430
SideEff3      Trend       0.1228     0.6665
SideEff4      Trend       0.0240     0.1555
SideEff5      Trend       0.0806     0.2420
SideEff6      Trend       0.1139     0.6325
SideEff7      Trend       0.0173     0.0995
SideEff8      Trend       0.1006     0.5205
SideEff9      Trend       0.3337     0.9430
SideEff10     Trend       0.3536     0.9735
```

While the original *p*-values suggested that at least two of the side effects had a significant dose-response at the 5% significance level, correcting for the multiple tests by permutation means reveals no significant differences at that level.

Summary

In this chapter, you learned how to combine the results of multiple, simultaneous analyses while controlling either the family-wise error rate or the false discovery rate.

To Learn More

Blair et al. (1996), Troendle (1995), and Westfall and Young (1993) expand on the use of permutation methods to analyze multiple hypotheses; also, see the earlier work of Shuster and Boyett (1979), Ingenbleek (1981), and Petrondas and Gabriel (1983).

Simultaneous comparisons in contingency tables are studied by Passing (1984).

Methods for determining the approximate minimal sample size for multiple comparisons may be gleaned from Dobbin and Simon (2005) and Lin et al. (2001).

9

Model Building

In this and the next chapter, we will be concerned with using the values of one or more predictor variables to predict the value of a dependent variable. In this chapter, we focus on the use of regression for prediction and show how the bootstrap and permutation methods may be used to estimate the values of model parameters, to test their significance, to estimate prediction errors, and to aid in validating the resulting model.

Regression Models

Regression models attempt to describe in quantitative terms the relationship between a dependent variable whose behavior we wish to explain or predict, in terms of one or more other variables, and its predictors. For example, if we feel that a snake's weight (W) is proportional to its length (S) and the square of its thickness (T), $W = a + bS + cT^2$, we would wish to estimate the *coefficients* a, b, and c.

The relationship may be *linear* in the unknown coefficients, as in the previous example, or it may be *nonlinear*, as in the logistic relationship between the probability of winning a contract, p, and the amount, A, that one is willing to bid, $p = (1 + ae^{-bA})^{-1}$.

It may involve a *single predictor*, as in the previous example, or *multiple predictors*, as in the case of the snake's weight.

The values of the predictors may be *fixed*, as in an experimental design, or they may be *random* samples from a larger population.

Regardless of how many predictors we employ, there always appears to be a part of the relationship that we cannot explain, that is due to other factors that we've neglected, or that may be attributed to experimental or observer error. The relationship between the dependent variable Y and its predictors X_1, X_2, ... is perhaps best written as a mixture of a *deterministic* component consisting of a known function of the predictors and a *stochastic* or random component.

Figures 9.1 and 9.2 illustrate the distinction between a strictly deterministic relationship and one with a random component.

The random component may have a *normal distribution*, a *Poisson distribution*, or some other not yet tabulated distribution.

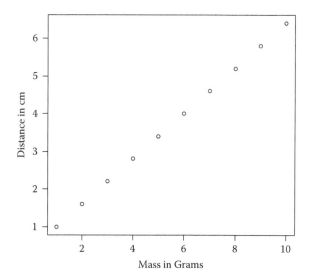

FIGURE 9.1
Hooke's law.

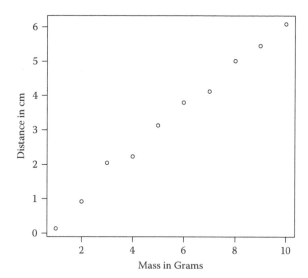

FIGURE 9.2
Student measurements.

   The method we employ to determine whether potential predictors have a statistically significant effect upon the dependent variable and to obtain confidence intervals for their coefficients will depend upon all the preceding distinctions.

As shown in Section 9.3, we may always make use of the bootstrap, though whether we bootstrap the residual errors or the observations will depend upon whether the predictors take fixed values or represent a random sample from a larger population.

As shown in Section 9.2, if the relationship involves a single predictor, then the more powerful permutation method may be employed.

Of course, if we can be certain the experimental errors have a normal, binomial, exponential, or Poisson distribution, then we may make use of parametric methods.

Bivariate Dependence

Before we begin to construct a model, we ought to ask whether the apparent association between Y, the variable we are trying to predict, and X, the predictor, is statistically significant. The Pearson correlation statistic is

$$\rho = \frac{Cov(XY)}{\sigma_X \sigma_Y}$$

where σ_X and σ_Y denote the standard deviations of the samples of X and Y, respectively, and the covariance of the two samples when the two variables X and Y are observed simultaneously is

$$Cov(XY) = \Sigma_{i=1}^{n}(x_i - \bar{x})(y_i - \bar{y})/(n-1)$$

If X and Y are independent and uncorrelated, then $\rho = 0$. If X and Y are totally and positively dependent, for example, if $X = 2Y$, then $\rho = 1$. In most cases of dependence

$$0 < |\rho| < 1$$

Good (2009) found that the Pearson correlation is highly robust against deviations from normality. In most practical applications, the use of resampling methods to determine its statistical significance is unnecessary.

Applying the Permutation Test

Whether and how one can test the significance of regression coefficients by permutation methods depends on the number of predictors.

Models with a Single Predictor

Seldom can we be sure that the errors in our observations have the normal distribution; to test whether the coefficient β in a model with a single predictor is significantly different from zero, we may make use of its permutation distribution.

In the following example, programmed in R, we first obtain coefficients for
a least absolute deviation regression,* then compute the permutation distribution of the slope to see if it is different from zero.

```
#obtain LAD regression coefficients and test slope to see if
greater than zero
library("quantreg")
Predictor =c (289,391,482,358,365,561,339,479,500,160,319,331)
Hopes = c(235,355,475,275,345,522,315,399,441,158,305,225)
N=400
f = coef(rq(formula = Hopes ~ Predictor))
   names(f)=NULL
   stat0=f[2]
   cnt=0
   for(i in 1:N){
      guestP=sample(Predictor)
      fp= coef(rq(formula = Hopes ~ Predictor))
      names(fp)=NULL
      if (fp[2] > = stat0) cnt=cnt+1
   }
return (cnt/N)
}
```

Comparing Two Regression Lines

A question that often arises in practice is whether two regression lines based
on two sets of independent observations have the same slope.† Suppose we
can assume

$$y_{ij} = a_i + bx_{ij} + \varepsilon_{ij} \text{ for } i = 1, 2; j = 1, ..., n_i$$

where the errors $\{\varepsilon_{ij}\}$ are exchangeable; then

$$\bar{y}_{i.} = a_i + b\bar{x}_{i.} + \bar{\varepsilon}_{i.}$$

---

* Our glossary of statistical terms contains definitions of the various regression methods.
† This is quite different from the case of repeated measures where a series of dependent observations are made on the same subjects (Good, 2005b, pp. 181–184) over a
period of time.

Define

$$y' = \frac{1}{2}(\bar{y}_1 - \bar{y}_2); \ x' = \frac{1}{2}(\bar{x}_1 - \bar{x}_2); \ \varepsilon' = \frac{1}{2}(\bar{\varepsilon}_1 - \bar{\varepsilon}_2); \ a' = \frac{1}{2}(a_1 + a_2).$$

Define

$$y'_{1i} = y_{1i} - y' \text{ for } i = 1 \text{ to } n_1 \text{ and } y'_{2i} = y_{2i} + y' \text{ for } i = 1 \text{ to } n_2.$$

Define

$$x'_{1i} = x_{1i} - x' \text{ for } i = 1 \text{ to } n_1 \text{ and } x'_{2i} = x_{2i} + x' \text{ for } i = 1 \text{ to } n_2.$$

Then

$$y'_{ij} = a' + bx'_{ij} + \varepsilon'_{ij} \text{ for } i = 1, 2; j = 1, \ldots, n_i$$

Two cases arise. If the original values of the predictors were the same for both sets of observations, that is, if $x_{1j} = x_{2j}$ for all j, then the errors $\{\varepsilon'_{ij}\}$ are drawn from the same distribution and we can apply the method of matched pairs, as in the "Other Two Sample Comparisons" section in Chapter 4. Otherwise, we need to proceed as follows: First, estimate the two parameters a' and b by least squares means. Use them to derive the transformed observations $\{y'_{ij}\}$. Then test the hypothesis that $b_1 = b_2$ using a two-sample comparison, as in the "A Distribution-Free Test" section. If the original errors were from a symmetric distribution and were exchangeable, then the transformed errors are exchangeable and this test is exact.

Alternately, we may know two curves are parallel but suspect they are not coincident. This problem is similar to that of the two-sample comparison, the difference being that we wish to increase the power of the test by correcting for the effects of various covariates. This problem also serves to illustrate some of the major differences between the permutation and the bootstrap approach.

Our solution by resampling methods requires us to take observations from the two populations for the same set of values of the covariates $\{X_i; i = 1, \ldots, n\}$. Given that,

$$Y_{1i} = a_1 + f[X_i] + \eta_{1i} \text{ for } I = 1, \ldots, n$$

$$Y_{2i} = a_2 + f[X_i] + \eta_{2i} \text{ for } I = 1, \ldots, n$$

where the $\{\eta_{ji}\}$ are independent random values each of whose expected value is zero.

To test the null hypothesis that $a_1 = a_2$, our statistic for the permutation test is that for matched pairs, $S = \Sigma(Y_{1i} - Y_{2i})$, where we perform n independent permutations, one for each pair. The permutation test is exact if we can assume the $\{\eta_{ji}\}$ are identically distributed.

The statistic for the bootstrap derived by Hall and Hart (1990) is

$$S = \left[\sum_{j=0}^{n-1}\left(\sum_{i=j+1}^{j+m} D_i\right)^2\right]\left[n\sum_{i=1}^{n-1}\left(D_{i+1} - D_i\right)^2\right]^{-1}$$

where

$$D_i = Y_{1i} - Y_{2i} - n^{-1}\sum_{i=1}^{n}(Y_{1i} - Y_{2i}) \text{ for } 1 \leq i \leq n$$

$$D_i = D_{i-n} \text{ for } n+1 \leq i \leq n+m$$

The complexities of this statistic are occasioned by the need to studentize to obtain asymptotically exact significance levels, and the introduction of m is necessary to prove that the results are asymptotically exact. The bootstrap test requires no additional assumption beyond the original one of the independence of the errors $\{\eta_{ji}\}$.

Multipredictor Regression

When more than one variable is used as a predictor, permutation tests for the individual regression coefficients are not exact. For suppose $y = \alpha + \beta_{x_1|x_2}x_1 + \beta_{x_2|x_1}x_2 + \varepsilon$ and we wish to test the hypothesis that $\beta_{x_2|x_1} = 0$ after correcting for the effect of the covariate x_1. Our test statistic is

$$\sum_{i<j} {}_{ijk}y \; {}_{ijk}\pi(x_2), \text{ where}$$

$${}_{ijk}y = \begin{bmatrix} 1 & 1 & 1 \\ y_i & y_j & y_k \\ x_{1i} & x_{1j} & x_{1k} \end{bmatrix}$$

and the permutations π are obtained by holding the couples (y_i, x_{1i}) fixed and randomizing with respect to x_2. If we knew what the relationship was between the dependent variable Y and the remaining predictors $X_2, X3, \ldots$ exactly, that is, if we knew what $\beta_{x_2|x_1}$ was and so needn't estimate it, we could obtain an exact test of the partial regression coefficient $\beta_{x_1|x_2}$ of Y on X_1 with ease. But because we must estimate all coefficients, the ones we are not interested in as well as the ones we are, these residuals are not exchangeable and the resulting test is not exact. Nor does this method of shuffling hold constant the collinearity between X_1 and X_2; this is a violation of the ancillary principle (see Welch, 1990). The resulting estimates of the coefficients will not vary as much in repeated samples as the original estimate causing the actual type I error to be larger than the declared significance level.

In an attempt to circumvent these difficulties, Tang et al. (2009) propose a permutation test that permutes treatment assignments for entire observations. Their technique mirrors the design of the study, with adjustment for nonresponders and baseline covariates. Specifically, they permute the randomization indicators according to the blocked randomization design, and then evaluate the intervention effect following the procedures in the original data analysis protocol, including weighting and covariate adjustments.

Consider a regression function: $E(Y) = g(\beta_0 + \beta_1 X + \beta_2 Z)$, where Z is the treatment indicator (1 = intervention, 0 = control), X is the vector of covariates, and g denotes a link function, with the identity link function appropriate for a linear model and the inverse of the logit function appropriate for a logistic regression model. The vector X could include fixed effects for blocks, or one can proceed with separate block effects.

The intervention effect is given by the regression coefficient β_2; the significance of the intervention effect is tested with the null hypothesis H_0: $\beta_2 = 0$ vs. H_a: $\beta_2 \neq 0$. The ratio $T = \beta_2/se[\beta_2]$ is used as the test statistic.

The null distribution of T for testing the hypothesis H_0 is simulated by rerandomizing the treatment assignment according to the original randomization protocol while keeping outcomes and covariates as observed. For a blocked randomized study with attrition weighting, follow this three-step procedure:

Step 1: Compute the test statistic for the actual data observed:
- Select an attrition weighting model.
- Fit the attrition weighting model.
- Derive attrition weights using the model fitted in step 1b.
- Fit the regression model using linear or logistic regression to obtain the test statistic T for the null hypothesis.

Step 2: Estimate the null distribution for the test statistic T with N replicates of the three substeps below:
- Rerandomize treatment assignment within each block.
- Rederive the attrition weights using the methods in step 1a–c.
- Refit the regression model using linear or logistic regression with attrition weights rederived in step 2b to rederive the test statistic T.

Step 3: Derive the empirical p-value for the test statistic T from the null distribution based on the permutation distribution obtained in step 2, i.e., $p = (M + 1)/(N + 1)$, where M denotes the number of replicates for which the test statistic T obtained in the permutation procedure in step 2 is equal to or greater than (in absolute value) the observed value of T obtained in step 1, and N denotes the total number of replicates. The addition of 1 in both the numerator and denominator represents the observed test statistic for the original data, which must be considered as one of the realizations of the permutation distribution.

Adaptive Regression

The parametric as well as the permutation tests of the various coefficients are not independent of one another. Thus, the significant levels of all the tests are suspect. The adaptive methodology due to O'Gorman (2006) provides tests for subsets of the coefficients that are independent of the values of the remaining predictors. His test is almost exact, with samples of 25 or more observations and 5 predictors. His method, described in what follows, is quite complicated. Fortunately, a SAS macro for executing his tests is available at the author's website (www.math.niu.edu/~ogorman/).

Suppose the complete model is given by $Y = X_R\beta_R + X_A\beta_A + \varepsilon$, where Y is an $n \times 1$ vector of observations, X_R an $n \times r$ matrix, β_R an $r \times 1$ vector of parameters, X_A an $n \times q$ matrix, β_A a $q \times 1$ vector of parameters, and ε an $n \times 1$ vector of errors, and we wish to test only a subset of the regression coefficients, H_0: $\beta_A = 0$ vs. H_a: $\beta_A \neq 0$.

To maximize power, use adaptive weightings. Begin by computing the studentized deleted residuals from the reduced model $Y = X_R\beta_R + \varepsilon$. Remove one observation at a time from the regression and compute the difference between the observed value and the predicted value based on the remaining $(n - 1)$ observations. Calculate the studentized deleted residuals as

$$d_i = e_i \sqrt{\frac{n-r-1}{SSE_R(1-h_{ii})-e_i^2}}$$

for $i = 1, \ldots, n$, where e_i is the ordinary residual from the reduced model, h_{ii} is the ith diagonal element of $X_R(X'_R X_R)^{-1}X_R$, and SSE_R is the usual sum of the squared residuals from the reduced model, all from the regression based on the n observations. Note that the studentized residuals are not independent.

To reduce the effects of nonnormality on the type I error, we smooth the cumulative distribution function (cdf) of the studentized deleted residuals using a normal kernel with a bandwidth of $h = 1.587 \, \hat{\sigma} n^{-1/3}$. Let $D = \{d1, \ldots, dn\}$. The smoothed cdf at point d over the set D is computed as

$$\hat{F}_h(d;D) = \frac{1}{n}\sum_{i=1}^{n} \Phi[(d-d_i)/h]$$

where $\Phi(\cdot)$ is the cdf of the standard normal distribution.

Center the studentized deleted residuals by subtracting the estimated median $\hat{d}$, $F_h(\hat{d}; D) = 0.5$. The centered studentized deleted residuals are calculated as $d_{c,i} = d_i - \hat{d}$. Let $D_c = \{d_{c,1}, \ldots, d_{c,n}\}$ and let $T_v(\cdot)$ be the cdf of the t distribution with v degrees of freedom.

Weight the individual observations, $w_i = T_{n-r-1}^{-1}[\hat{F}_h(d_{c,i}; D_c)] / d_{c,i}$ for $i = 1, \ldots,$ n., and use the weights w_i as the diagonal elements in a weighting matrix W that has zero off-diagonal elements. Compute the weighted least squares regression by premultiplying both sides of the complete model by W to obtain $WY = WX_R\beta_R + WX_A\beta_A + W\varepsilon$. Use ordinary least squares to compute SSE for this model, denoted by SSE^*_C. Similarly, premultiply the reduced model by W and compute SSE^*_R.

To obtain a permutation test of our H_0: $\beta_A = 0$ vs. Ha: $\beta_A \neq 0$, proceed as follows:

- Regress Y on X_R alone to obtain the vector of predicted values $Y^\wedge$ and the residual vector e.
- Compute the test statistic $F^* = \dfrac{(SSE^*_R - SSE^*_C)/q}{SSE^*_C/(n-r-q)}$ for the unpermuted data.
- Permute the rows of e and denote the ith permutation of the elements in e as e_i. Let $Y^i = Y^\wedge + e_i$. For the ith permutation, compute the adaptive weights and the test statistic F^* from the adaptively weighted regression of Y^i on X_R and X_A, as described above.

Repeat step 3 400 plus times to obtain an estimate of the p-value as described in Chapter 4.

Applying the Bootstrap

The R and Stata codes that follow can be quickly adapted for use with multiple predictors.

```
#obtain bootstrap confidence intervals for LAD regression
coefficients
library("quantreg")
Predictor =c (289,391,482,358,365,561,339,479,500,160,319,331)
Hopes = c(235,355,475,275,345,522,315,399,441,158,305,225)
n = length(Predictor)
data=cbind(Predictor,Hopes)
#set number of bootstrap samples
N =400
stat = numeric(N) #create a vector in which to store the results
for(i in 1:N){
```

```
    ind=sample(n,n, replace=T)
    guestP= data[ind,]
    fp= coef(lq(formula = Hopes ~ guestP))
    stat[i]= fp[2]
}
quantile(stat,prob=c(0.05,0.95))
```

Stata

```
permute Hopes "regress Hopes Predictor" _b, reps(400) left
command: regress Hopes Predictor
statistics: b_Predictor = _b[Predictor]
b_cons = _b[_cons]
permute var: Hopes
```

```
Monte Carlo permutation statistics Number of obs =12
Replications =400
```

```
T | T(obs) c n p=c/n [95% Conf.Interval]
b_Predictor | .9393529 400 400 1.0000 .99082021
b_cons |-20.55001 0 400 0.0000 0.0091798
```

```
Note: confidence intervals are with respect to p=c/n
Note: c = #{T <= T(obs)}
```

```
. bootstrap "regress Hopes Predictor" _b, reps(400) nonormal
nopercentile
command: regress Hopes Predictor
statistics: b_Predictor = _b[Predictor]
b_cons = _b[_cons]
```

```
Bootstrap statistics Number of obs =12
Replications =400
```

```
Variable | Reps Observed Bias Std. Err. [95% Conf. Interval]
b_Predictor | 400 .9393529 .0072056 .0921974 .8209208
1.190321(BC)
b_cons | 400-20.55001-3.715155 38.81548-140.0799 24.55453(BC)
```

The preceding code can be used to test for significance by replacing the single predictor regression commands with multivariable ones. In Stata, for example, one merely needs to write

```
. bootstrap "regress Hopes Predictor Weather" _b, reps(400)
nonormal nopercentile
```

If our results are to apply to a population of which the data are only a random sample, then as we did above, we resample complete observations consisting of both predictors and the response variable. For fixed values of the predictors, on the other hand, bootstrapping is based on repeated sampling from the residuals of the regression model. Typically, bootstrapping for fixed predictors yields somewhat narrower confidence intervals than bootstrapping for random predictors. Consider the 95% bootstrap confidence intervals for lmg shares in the Swiss data set:

```
> fixedlmg <-booteval.relimp(boot.relimp(linmod, b = 1000,
fixed = TRUE),
+ bty = "perc," level = 0.95)
> randomlmg <-booteval.relimp(boot.relimp(linmod, b = 1000),
bty = "perc,"
+ level = 0.95)
> output <-rbind(fixedlmg$lmg.lower, fixedlmg$lmg.upper,
+ randomlmg$lmg.lower, randomlmg$lmg.upper)
> output <-as.matrix(t(output))
> colnames(output) <-c("fixed.lower," "fixed.upper,"
"random.lower,"
+ "random.upper")
> rownames(output) <-c(fixedlmg$namen[2:6])
> output
```

	fixed lower	fixed upper	random lower	random upper
Agriculture	0.047	0.088	0.031	0.113
Examination	0.119	0.264	0.089	0.288
Education	0.169	0.377	0.069	0.384
Catholic	0.045	0.196	0.037	0.239
Infant.Mort.	0.035	0.218	0.029	0.238

Building a Model

Overfitting is a risk associated with all model building efforts; in an effort to fit each and every value in the data at hand, the resultant model proves of little predictive value when confronted with new data. The noisier the data and the greater the proportion of outlying values, the more likely one is to overfit.

The RANSAC algorithm due to Fischler and Bolles (1981) uses as small of an initial data set as feasible, then attempts to enlarge it through the addition of compatible data. We proceed as follows.

Suppose we are given a model that requires a minimum of n data points to determine its free parameters (for example, $n = 2$ for a straight line, $n = 3$ for a circle), and a sample S of size $N > n$. Set $k = 1$.

Select a subsample S_k of size n from S at random and use it to derive a model M_1. Add all points of the original sample to the subsample that are within some predetermined error tolerance of M_1. Call the combined sub-sample S^*_k.

Case 1: Size$(S^*_k)/N$ is greater than some predetermined threshold T. Use the data in S^*_k to determine a new model M and quit.

Case 2: Size$(S^*_k)/N$ is less than some predetermined threshold T. If the number of random subsamples k is less than some predetermined value K, set $k = k + 1$ and repeat step 1. Otherwise consider adding an additional parameter to the model.

This method has seen a wide variety of applications, particularly in image processing and pattern matching; see, for example, Nistér et al. (2006) and Stewart et al. (2003). C++ routines for use in pattern matching may be downloaded from http://public.kitware.com/vxl/doc/release/contrib/rpl/rgrl/html/index.html

Limitations of the Bootstrap

The bootstrap procedures described in this chapter and in Chapter 2 cannot be relied on when there are too few observations in the original sample. Increasing the number of bootstrap samples will not remedy the situation when there is inadequate information to begin with. SAS reports at http://ftp.sas.com/techsup/download/stat/jackboot.html that when attempts were made to find a 95% confidence interval for R^2 in a linear regression with 20 observations and 10 predictors, the bootstrap distribution was not even close to the true sampling distribution. The bootstrap BCa interval was extremely short and did not contain the true value.

For comparisons of mean values or percentiles close to the mean, a sample of at least 25 observations is recommended. For percentiles further from the median, even larger samples are recommended. If we need n^2 observations to ensure that the distribution of one predictor in the sample is sufficiently close to its distribution in the population, then if we have k different independent predictor variables, we will need $(n^2)^k$ multivariate observations in the original sample to ensure similar accuracy.

Fortunately, in most practical cases, our predictor variables are not independent but correlated, and we can ensure a good fit between the sample distribution and the population distribution with a much smaller sample, though still of order n^{2j}, where $1 < j < k$.

Prediction Error

To estimate the accuracy of our model, two approaches are possible: cross-validation and the double bootstrap.

Cross-Validation

The data are divided randomly into K equal-sized groups. For each group, we fit the generalized linear model to the data remaining after omitting that group. The prediction error is based on the difference between the observed responses in the group that was omitted from the fit and the prediction made by the fitted models for those observations.

In R, we install the boot package, then make use of the function cv.glm(data, glmfit, cost, K), where data is the matrix or data frame containing the data, glmfit contains the results of a generalized linear model fitted to the data, and $K \geq 1$ is the number of equal-sized groups. The default for cost is the average squared error function, but you may substitute any other loss function.

Double Bootstrap

When we estimate a population mean using the sample mean, the result is unbiased; that is, the mean of the means of all possible samples taken from a population is the population mean. In the majority of cases, including the preceding example, estimates of error based on bootstrap samples are biased and tend to underestimate the actual error. If we can estimate this bias by bootstrapping a second time, we can improve on our original estimate.

This bias, which Efron and Tibshirani (1993) call the optimism, is the difference in the average values $E[w, F] - E[w, F^b]$, which we estimate as before by using the observations in place of the population and the bootstrap samples in place of the observations,

$$E[w^*, F^b] - E[w, F^{b*}] = \Sigma^B \Sigma_i \{L(m_i, m[x_i, w^*]] - L[m_i^*, m[x_i^*, w^*])\}/nB$$

$L(m_i, m[x_i, w^*])$ uses the observations from the original sample and the coefficients derived from the bootstrap; in contrast, $L(mi^*, m[xi^*,w^*])$ uses the bootstrap observations and the coefficients derived from the bootstrap. Our corrected estimate of error is the apparent error rate plus the optimism, or

$$E[w, F^b] + E[w^*, F^b] - E[w, F^{b*}]$$

Validation

How can we be confident of our ability to predict the future when all we have to work with is the past? Before applying a model we need to *validate* it. Ideally, we would have some independent source of verification we could draw on. But in most instances, we will have no alternative but to make use of the same data to validate the model that we used to develop it. To do so, we need to either develop a *metric* and determine its permutation distribution, bootstrap, or apply some other form of *cross-validation*. In what follows we consider each of these approaches in turn.

Metrics

To distinguish between close and distant, between a good-fitting model and a poor one, we first need a metric, and then a range of typical values for that metric. A metric m defined on two points x and y has the following properties:

$$m(x, y) \geq 0,$$

$$m(x, x) = 0,$$

$$m(x, y) \leq m(x, z) + m(z, y).$$

These properties are possessed by the common ways in which we measure distance. The third property of a metric, for example, simply restates that the shortest distance between two points is a straight line.

A good example is the standard Euclidian metric used to measure the distance between two points x and y whose coordinates in three dimensions are (x_1, x_2, x_3) and (y_1, y_2, y_3):

$$\sqrt{(x_1 - y_1)^2 + (x_2 - y_2)^2 + (x_3 - y_3)^2}$$

This metric can be applied even when x_i is not a coordinate in space but the value of some variable like blood pressure or laminar flow or return on equity. Hotelling's T^2, defined in Chapter 5, is an example of such a metric.

As in Chapter 5, the next step is to establish the range of possible values that such a metric might take by chance, using either the bootstrap or permutations of the combined set of theoretical and observed values. If the value of this metric for the original observations is an extreme one, we should reject our model and begin again.

Nearest Neighbors

Suppose we have a set of N observations and a second set of N simulated values. For each point in the set of theoretical values, set 1, record the number of points among its three nearest neighbors (nearest in terms of the Euclidian metric) that are also in set 1; sum the number of points so recorded. Call this sum S_0.

To determine whether S_0 is an extreme value, compute S for rearrangements of the combined group of theoretical and observed values. In all but the original and one other arrangement, set 1 will consist of a mixture of observed and theoretical values. If these values are significantly different from one another, fewer nearest neighbors will be in the same set and S will be smaller than S_0.

Goodness of Fit

A metric can always be derived. Before developing a model, divide the data at random into two parts, one that will be used for model development and estimation, and the other for validation. Our goodness-of-fit metric is

$$G = \frac{\sum_{k\in\{\text{validation}\}} (Y_{\text{observed}} - Y_{\text{predicted}})^2}{\sum_{k\in\{\text{estimation}\}} (Y_{\text{observed}} - Y_{\text{predicted}})^2}$$

where the summation in the numerator is taken over all the observations in the validation data set, and the summation in the denominator is taken over all the observations in the estimation data set.

For two reasons, this ratio will almost always be larger than unity, and it would not be appropriate to assess its value via the F-distribution:

1. The estimation data set, not the validation set, was used to choose the variables that went into the model and to decide whether to use the original values of the variables or to employ some sort of transformation.
2. The estimation data set, not the validation set, was used to estimate the values of the model coefficients.

Divide the original data set into two parts at random a second time, but use the estimation set only to calculate the values of the coefficients. Use the same model you used before; that is, if log[X] was used in the original model, use log[X] in this new one. Compute G a second time.

Repeat this resampling process several hundred times. If the original model is appropriate for prediction purposes, it will provide a relatively good fit to most of the data sets; if not, the goodness-of-fit statistic for our original estimation set will be among the largest of the values, since its denominator will be among the smallest.

Warning: Goodness of fit to the data you used to create the model proves nothing; in fact, this approach generally yields to *overfitting* so that the resultant model has less predictive value when applied to new data. In the words of John von Neumann, "With four parameters I can fit an elephant and with five I can make him wiggle his trunk."

GUIDELINES FOR MODEL BUILDING

Your objective in modeling is prediction, not goodness of fit:

- Use the minimum number of predictors—overfitted models are numerically unstable.
- Use automated methods only to produce several apparently good models that can be investigated further.
- Include only predictors with which a plausible cause-and-effect relationship can be established.
- Replace multiple highly correlated (collinear) predictors with an average or some other linear combination. This procedure recommends itself when a coefficient that should be positive (negative) has the opposite sign.
- When investigating models for several similar products, select the model that fits all the products reasonably well, rather than trying to find the best model for each product. (Any exception should be justified on a cause-and-effect basis.)

Using the Bootstrap for Model Validation

If choosing the correct functional form of a model in the case of a single variable presents difficulties, consider that in the case of k variables, there are k linear terms (should we use logarithms? should we add polynomial terms?) and $k(k-1)$ first-order cross-products of the form $X_i X_k$. Should we include any of the $k(k-1)(k-2)$ second-order cross products?

The obvious solution (similar to belling the cat) is to confine the task to just the important variables, the ones that bear a causal relationship to the variable we are trying to predict or the hypothesis we are proposing to test. The bootstrap can help.

In the late 1970s, Peter Gregory observed 155 chronic hepatitis patients at Stanford Hospital and made 20 observations on each one summarizing medical histories, physical examinations, x-rays, liver function tests, and biopsies. Gail Gong (1986) constructed a logistic regression model based on his observations; the object of the model was to identify patients at high risk.

Gong's logistic regression models were constructed in two stages. At the first stage, each of the 19 explanatory variables was evaluated on a univariate basis. Thirteen of these variables proved significant at the 5% level when applied to the original data. A forward multiple regression was applied to these 13 variables, and 4 were selected for use in the predictor equation.

She then proceeded to take several hundred bootstrap samples of the 155 patients. Each bootstrap sample was obtained by drawing with replacement from the 155 vectors of observations. Thus, one patient's observations might be missing completely from the bootstrap sample, while another's might be duplicated two or even three or more times.

Computer programs that perform the bootstrap typically draw a sample with replacement from the one-dimensional list of patient's names, then construct a multidimensional matrix of the associated data. In Gong's (1986) case, the size of the bootstrap sample, m, was the same as the original sample size n. In cases where the number of variables is an order of magnitude greater than the number of observation vectors, as with microarrays and images, it is preferable to take $m < n$, a procedure known as subsampling.

When she took bootstrap samples of the 155 patients, the R^2 values of the final models associated with each bootstrap sample varied widely. Not reported in her article, but far more important, is that while two of the original four predictor variables always appeared in the final model derived from a bootstrap sample of the patients, five other variables were incorporated in only some of the models. Surprisingly, only four of the predictors were common to each of the 100 models Gong (1986) developed. There are two conclusions to be drawn from this. First, any regression model (linear or nonlinear) should be regarded with a great deal of skepticism. Second, the bootstrap is invaluable for differentiating among the essential and inessential components of a model.

R Code

```
G=glm(V1~-1+V2+...,,hepdata,na.action=na.omit)
step(G)
le=length(hepdata$V1)
mmat=data.matrix(hepdata)
bmat=mmat
for (j in 1:100){
   bdex=sample(1:le,replace=T)
   for (i in 1:le) bmat[i,]=mmat[bdex[i],]
   bdata=as.data.frame(bmat)
   G=glm(V1~-1+V2+...,,bdata,na.action=na.omit)
   step(G)
}
```

Cross-Validation

The bootstrap is only one of several possible methods of cross-validation. The following also are in general use:

K-fold, in which we subdivide the data into K roughly equal-sized parts, then repeat the modeling process K times, leaving one section out each time for validation purposes.

Leave one out, an extreme example of *K*-fold, in which we subdivide into as many parts as there are observations. We leave one observation out of our classification procedure, and use the remaining $n - 1$ observations as a training set. Repeating this procedure n times, omitting a different observation each time, we arrive at a figure for the number and percentage of observations classified correctly. A method that requires this much computation would have been unthinkable before the advent of inexpensive, readily available high-speed computers. Today, at worst, we need to step out for a cup of coffee while our desktop completes its efforts.

Jackknife, an obvious generalization of the leave-one-out approach, where the number left out can range from one observation to half the sample.

Delete–d, where we set aside a random percentage d of the observations for validation purposes, use the remaining 100–d% as a training set, then average over 100 to 200 such independent random samples. The goodness-of-fit statistic is an example of delete–50.

Jiang and Simon (2007) compare the various cross-validation methods when used for estimating the prediction error in microarray classification. Vickers et al. (2008) found 10-fold cross-validation methods superior to the bootstrap when correcting for overfitting. As we shall see in the next chapter, 10-fold methods are the standard in building decision trees.

Summary

In this chapter, you reviewed the steps in model development and considered some of the possible caveats. You learned a number of resampling methods for ascertaining the statistical significance of model coefficients and for deriving the associated confidence intervals. You learned it is essential to distinguish between goodness of fit and prediction, and were provided with a variety of cross-validation resampling methods to tie the two together.

To Learn More

Regression is a rich and complex topic. Ryan (1997) provides a review of the many methods of estimation. Mosteller and Tukey (1977) and Good and

Hardin (2009) document the many pitfalls. Cardot et al. (2007) have applied permutation methods to nonlinear equations of the form $m(t, x) = E[Y(t)|X = x]$.

Attempts at formulating exact permutation tests for the coefficients of multipredictor regression are summarized by Anderson and Legendre (1999). Good (2000, pp. 127–130) cites the lack of exchangeability as responsible for the deviations from the declared significance level. Despite these caveats, Cade and Richards (2006) have applied permutation methods to multivariate quantile regression, and Luger (2005) to multivariate nonlinear regression.

Jiang and Simon (2007) and Binder and Schumacher (2008) have used bootstrap methods to estimate the prediction errors in microarray classification.

For an example of the bootstrap's application in a nonlinear regression, see Shimbukaro et al. (1984).

Karlin and Williams (1984) developed their own metric for use in a structured exploratory data analysis of familial traits. Mielke (1986) reviews alternate metrics. Marron (1987) and Hjorth (1994) provide a survey of parameter selection and cross-validation methods. Techniques for model validation also are reviewed in Shao and Tu (1995, pp. 306–313), who show that the delete–50% method, first proposed by Geisser (1975), is far superior to delete–1.

Exercises

1. Characterize the following relationships as positive linear, negative linear, or nonlinear:

   a. Distance from the top of a bathtub to the surface of the water as a function of time when you fill the tub, then pull the plug

   b. Same problem, only now you turn on the taps full blast

   c. Sales as a function of your advertising budget

   d. Blood pressure as you increase the dose of a blood-pressure-lowering medicine

   e. Electricity use as a function of the day of the year

   f. Number of bacteria at the site of infection as you increase the dose of an antibiotic

   g. Size of an untreated tumor over time

2. Which is the cause and which the effect?

   a. Overpopulation and poverty

   b. Highway speed limits and number of accidents

   c. Cases of typhus and water pollution

   d. Value of the euro in U.S. dollars

3. In the following table, breaks appear to be related to log dose in accordance with the formula breaks = $a + b$log[dose + 0.01].

   a. Estimate b.

   b. What does a represent? How would you estimate a?

   Micronuclei in Polychromatophilic Erythrocytes and Chromosome Alterations in Bone Marrow of CY-Treated Mice

Dose (mg/kg)	Number of Animals	Micronuclei per 200 Cells	Breaks per 25 Cells
0	4	0 0 0 0	0 1 1 2
5	5	1 1 1 4 5	0 1 2 3 5
20	4	0 0 0 4	3 5 7 7
80	5	2 3 5 11 20	6 7 8 9 9

4. Suppose you wanted to predict the sales of automobile parts; which observations would be critical? Number of automobiles in service? Number of automobiles under warranty? Number of automobiles more than a given number of years in service? M2, the total money supply that is readily available in cash, savings, and checking accounts, and so forth? Retail sales? Consumer confidence? Weather forecasts? Would your answer depend on whether you owned a dealership or an independent automobile parts chain? (By the way, in the studies Mark Kaiser and I completed, weather was the most significant variable. Go figure.)

5. Suppose you've collected the prices of a number of stocks over a period of time and have developed a model with which to predict their future behavior. How would you go about validating your model?

6. The State of Washington uses an audit recovery formula in which the percentage to be recovered (the overpayment) is expressed as a linear function of the amount of the claim. The slope of this line is close to zero. Sometimes, depending on the audit sample, the slope is positive, and sometimes it is negative. Can you explain why?

10

Classification

How can we tell whether the mushroom we've just found is edible or poisonous? Whether an incoming letter is spam? Or whether the stranger on the phone is a telemarketer—preferably before we pick up the receiver? In making these decisions, what factors should be considered? Which are most meaningful?

Is the mushroom poisonous?

Does it have a cap with gills?

Yes. Are the gills free from the stalk?

Yes. Is there a volva present?

Yes. Is the volva fibrillose to membraneous?

Yes. Is the cap pure white?

Stay away! The tempting fungi we're holding is an *Amanita virosa*, also known as a destroying angel.

In the previous example, we made an identification with just five questions. It takes 22 points of comparison for the FBI to match a fingerprint with one of the millions of prints on file in its electronic database. Again, a decision tree or key is involved in which the answers to a series of questions lead finally to a decision. Decision trees are at the basis of most spam filters and ad eaters; see, for example, http://www.rulequest.com/see5-examples.html#ADEATER.

What questions should we ask to help us make our decisions? Which answers will be most meaningful? In this chapter, you will learn how to construct decision trees and study their use in classification and as an alternative to regression in model development and the analysis of microarrays and complex images. You will also learn that the ambiguities inherent in stepwise regression are still present with decision trees, and that validation again is essential.

Cluster Analysis

When we have not yet established a classification scheme, *cluster analysis* may help us. Its purpose is to establish relationships or classifications among phenomena, for example, to define market segments, to find structural similarity among chemical compounds, and to find in-function among genes.

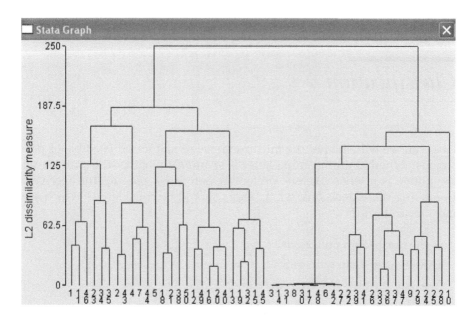

FIGURE 10.1a
Dendrogram of the original sample.

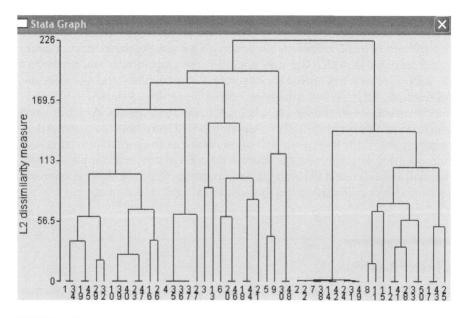

FIGURE 10.1b
Dendrogram of a bootstrap sample.

Just as the bootstrap may be used to validate regression models, least squares, quantile, or nonlinear, so it is also recommended for use in validating cluster analysis. See, for example, Figure 10.1a and b. Bootstrapping cluster analysis begins with creating a number of simulated data sets based on the statistical model. If ANOVA is the appropriate model, as proposed by Kerr and Churchill (2001), the bootstrap-simulated data sets would take the form

$$y_{ijk}^{*} = f(\vec{x}_k) + \varepsilon_{ijkg}^{*}$$

where $f(\vec{x}_k)$ is a linear function of the predictors whose coefficients are estimates from the original model fit. The error terms ε_{ijkg}^{*} are drawn *with replacement* from the studentized residuals of the original model fit. Repeat the clustering procedure on each simulated data set, $y^{*} \rightarrow \hat{r}^{*} \rightarrow \hat{C}^{*}$, to obtain a collection of bootstrap clusterings $\{\hat{C}^{*}\}$.

This approach is inadequate absent some measure of consistency. Dolnicar and Leisch (2010) recommend that one compute a cluster index, the Calinski-Harabas index (Milligan and Cooper, 1985), or the Rand index (Hubert and Arabie, 1985) for each bootstrap data set. As the number of clusters as well as the composition of those clusters will vary from bootstrap sample to bootstrap sample, only clusters that appear with some regularity should be the object of further investigation.

Zhang and Zhao (2000) propose the use of decision trees (the focus of the balance of this chapter) to be applied to a series of bootstrap samples taken from the original data. Their approach is similar to that of Gong's (1986) in that a majority rule consensus tree is obtained, showing clusters that only occur in a majority of the resampled trees.

Classification

Anderson (1935) recorded the sepal length, sepal width, petal length, and petal width of 150 iris plants. The data he collected may be downloaded from http://stat.bus.utk.edu/Stat579/iris.txt. The presence of at least two species of iris seems evident from the histogram of the data in Figure 10.2, though because of the overlap of the various subpopulations, it is difficult to be sure. Three species actually are present, as shown in Figure 10.3.

To develop a decision tree to help classify a newfound iris plant, begin by downloading a trial version of See5® from http://www.rulequest.com/download.html.

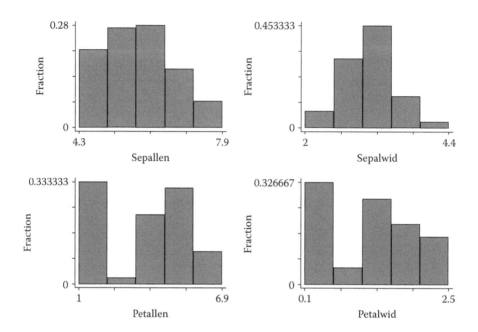

FIGURE 10.2
Sepal and petal measurements of 150 iris plants.

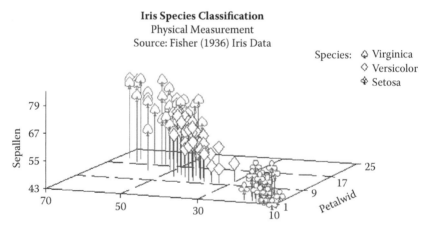

Petallen: petal length in mm.   Petalwid: petal width in mm.
Sepallen: sepal length in mm.   Sepal width not shown.

FIGURE 10.3
Representing three variables in two dimensions. Iris species representation derived with the help of SAS/Graph®.

Read 150 cases (5 attributes) from iris.data

Decision tree:

```
petal length <= 1.9: setosa (50)
petal length > 1.9:
:...petal width > 1.7: virginica (46/1)
   petal width <= 1.7:
   :...petal length <= 4.9: versicolor (48/1)
      petal length > 4.9: virginica (6/2)
```

FIGURE 10.4
See5 decision tree for iris classification.

TABLE 10.1

Learning Sample Classification Table

Actual Class	Predicted Class 1 2 3			Actual Total
Predicted total	50	54	46	150
Correct	1.00	0.98	0.90	0.96

Decision trees built by See5 are formed by binary splits of one of two types:

1. The subject of the investigation has a certain property, or it does not have that property.
2. The value of a property of the subject is less than or equal to a number k, or it is greater than k.

First, See5 ranks all the characteristics—sepal length, sepal width, petal length, and petal width—as to which is the most informative. In this example, it establishes that petal length is the most informative. And the most informative split is on the classification rule "petal length less than or equal to 1.9," in which case all such observations may be assigned to the class setosa. The second split is on the rule "petal width less than or equal to 1.7." See5 concludes the sepal measurements do not contribute significant additional information. See Figure 10.4 for the finished result.

As shown in Table 10.1, 96% are correctly classified! A moment's thought dulls the excitement; this is the same set of observations we used to develop our selection criteria, so naturally we were successful. The true test will be whether we can achieve similar success in classifying new plants whose species we don't know in advance.

Not having a set of yet-to-be-classified plants on hand, the classification and regression tree (CART®) method makes use of the existing data in a 10-fold cross-validation, as described in the "Goodness of Fit" section in Chapter 9. It divides the original group of 150 plants into 10 subgroups, each consisting

TABLE 10.2

Cross-Validation Classification

Class	Prior Prob.	Learning Sample			Cross-Validation		
		N	Misclassified	Cost	N	Misclassified	Cost
1	1/3	50	0	0.00	50	0	0.00
2	1/3	50	1	0.02	50	5	0.10
3	1/3	50	5	0.10	50	5	0.10
	1	150	6		150	10	

of 15 plants. Its binary tree-fitting procedure is applied 10 times in succession, each time omitting one of the 10 groups, and each time applying the tree it developed to the omitted group. As shown in Table 10.2, only 10 of the 150 plants, or 7.5%, are misclassified, a 92.5% success rate, suggesting that in the future, irises of unknown species would be classified successfully 92.5% of the time.

To obtain similar results using R commands, first download and install the "tree" library:

```
# Species must be a factor
iris.tr=tree(Species ~ SL+SW+PL+PW)
plot(iris.tr); text(iris.tr,srt=90)
```

The "Software for Creating Decision Trees" section contains a list of recommended software.

Decision Trees

In analyzing crop yield variability, Tittonell et al. (2008) found that CART (classification and regression tree) software helped to stratify variability into classes that reflected interactions between crop management and soil fertility.

Figure 10.5 depicts the model setup in CART for predicting low birth weight (low) on the basis of the mother's age and race, whether she smoked during pregnancy (smoke), had a history of hypertension (HT) or uterine irritability (UI), how often she visited a physician during the first trimester (FTV), and so forth.

Note that in contrast to *continuous* measurements like petal length and age, for which decision trees send a case left if the value of a split variable is less than a cutoff value, with categorical values like race and history (or no history) of hypertension, a case goes left if the split variable takes one of a specified set of values.

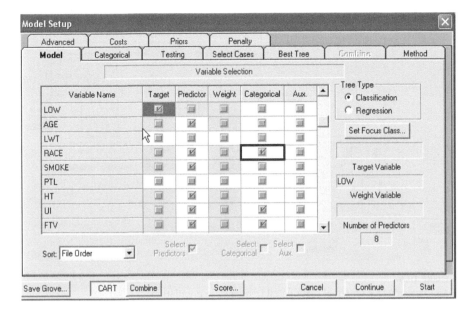

FIGURE 10.5
Model setup in CART for predicting low birth weight.

Refining the Model

Unless we specify the costs of misclassification for the different categories, CART treats them as if they are the same. Similarly, unless we specify the probabilities of the various outcomes, CART assumes that low and high birth weights are equally likely. The resulting decision tree with equal costs and equally likely outcomes is displayed in Figure 10.6.

If instead we specify that a higher cost is incurred should the tree predict a normal birth weight when the actual birth weight is low, Figure 10.7 depicts the decision tree.

The frequencies of the various classifications are seldom equal, nor are the sample frequencies necessarily representative of the proportions in the entire population. If we know what these frequencies are, we should specify them in order to reduce the overall error rate during all subsequent classification phases. The resulting decision tree is shown in Figure 10.8.

Decision Trees vs. Regression

Which method should we employ when we want to go beyond classification to making specific numerical predictions?

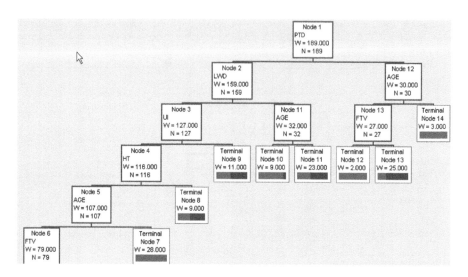

FIGURE 10.6
CART decision tree for equal costs, equal frequency in population.

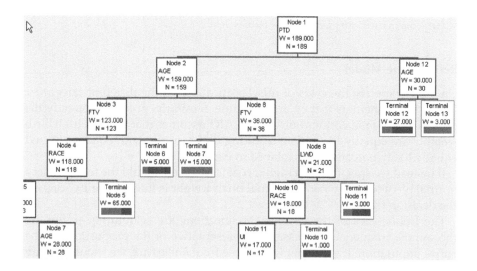

FIGURE 10.7
CART Decision tree for unequal costs, equal frequency in the population.

Recall that the basis of the regression techniques discussed in Chapter 9 is a desire to minimize the sum of the potential losses $\sum_i L(\hat{\theta}_i - \theta_i)$. With regression techniques, we are restricted to estimates that take the linear form $\hat{\theta}_i = AX_i$ and loss functions L that can be minimized by least squares or linear programming techniques. With decision trees, we are able to specify loss functions of arbitrary form. Moreover, if we have some prior knowledge

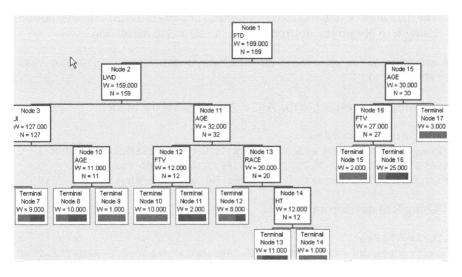

FIGURE 10.8
CART decision tree: Equal costs, unequal frequency of categories in population.

of the distribution of the various classifications in the population at large (as opposed to what may be a quite different distribution in the sample at hand), we can take advantage of this knowledge to obtain greater success in classification.

In this section, we develop two models for predicting home values, the first by regression means, the second via a CART decision tree. The data consists of the median value (MV) of owner-occupied homes in about 500 U.S. census tracts in the Boston area, along with several potential predictors, including:

CRIM   per capita crime rate by town

ZN   Proportion of residential land zoned for lots over 25,000 ft^2

INDUS Proportion of nonretail business acres per town

CHAS   Charles River dummy variable
         (= 1 if tract bounds river; 0 otherwise)

NOX   Nitric oxides concentration (parts per 10 million)

RM   Average number of rooms per dwelling

AGE   Proportion of owner-occupied units built prior to 1940

DIS   Weighted distances to five Boston employment centers

RAD   Index of accessibility to radial highways

TAX   Full-value property tax rate per $10,000

PT   Pupil-teacher ratio by town

LSTAT   Percent lower status of the population

The data may be downloaded from http://lib.stat.cmu.edu/datasets/boston.
Using R to obtain the desired results via stepwise regression:

```
summary(step(lm(MV~ CRIM+ZN+INDUS+CHAS+NOX+RM+AGE+DIS+RAD+TAX+
PT+B+LSTAT)))
```

and fitting with respect to the AIC criterion yields the model:

```
Step: AIC= 1585.76
MV = 36.3-0.11CRIM + 0.05ZN + 2.7CHAS + 17.38NOX + 3.80RM +
1.49DIS + 0.30RAD-0.01 TAX + 0.94PT + 0.009B-0.52LSTAT
With Multiple R-Squared: 0.7406, Adjusted R-squared: 0.7348
```

Using R to develop a tree

```
library("tree")
bos.tr=tree(MV~ CRIM+ZN+INDUS+CHAS+NOX+RM+AGE+DIS+RAD+TAX+PT+B
+LSTAT)
summary(bos.tr)
```

yields the following output:

```
Variables actually used in tree construction:
[1] "RM" "LSTAT" "DIS" "CRIM" "PT"
Number of terminal nodes: 9
Residual mean deviance: 13.55 = 6734 / 497
```

The command

```
plot(bos.tr); text(bos.tr, srt=90)
```

yields the tree displayed in Figure 10.9. Note that the same variables are
employed as were obtained via a stepwise regression.

The most important distinction between the two methods, regression and
decision trees, is that while the magnitude of the regression coefficients do not
lend themselves to cause-and-effect interpretation, the branches of the tree
suggest possibilities for further study. The decision tree provides a direct estimate of prediction error; the R^2 of regression requires further interpretation.

Which Predictors?

The cost of collecting data may vary from predictor to predictor. As our
potential predictors are interdependent, we may wish to consider what
would happen if we were to substitute variables. Table 10.3 orders the predictors in their order of importance should they be used in the absence of
other information.

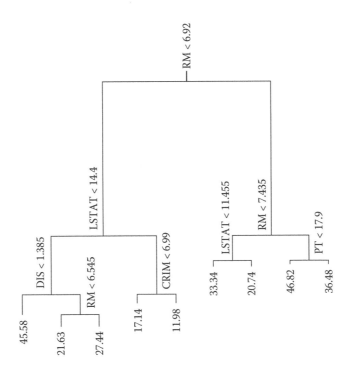

FIGURE 10.9
Regression tree for predicting median home value.

TABLE 10.3

Variable Selection in Order of Importance Using Ordinary Least Squares

Variable	LSAT	RM	DIS	NOX	PT	INDUS	TAX	AGE	CRIM	ZN	RAD
Score	100	90	29	26	25	23	20	15	13	11	5

Though crime rates (CRIM) figure early in the decision tree of Figure 10.9, information concerning crime rates may not be readily available. Figure 10.10 depicts the revised decision tree should we omit this variable.

Which Decision Tree Algorithm Is Best for Your Application?

Whenever I've tried to persuade researchers to make use of decision trees, their eyes glaze over. For them, the computer will always be where they send and receive email, consult Wikipedia and porn sites, and on rare occasions, apply some statistical routine they are familiar with and can access with a single key press.

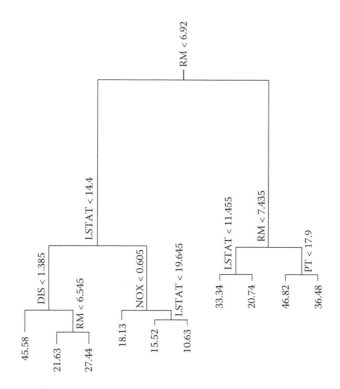

FIGURE 10.10
Decision tree omitting CRIM.

Alas, making effective use of decision trees is far more complicated. As shown in the preceding examples, one can access and make use of the methodology with a single key press. But to make the most effective use, considerable forethought is necessary.

Focus solely on predictors that bear some causal relationship to the outcomes of interest. Make use of any prior knowledge, such as the observed frequency of the various possible classifications.

Different tree algorithms yield different results. For example, Trujillano et al. (2009), attempting to classify critically ill patients in terms of their severity, found that CART made use of 5 variables and 8 decision rules; CHAID, 7 variables and 15 rules; and C4.5, 6 variables and 10 rules.

CHAID (Chi-squared Automatic Interaction Detector) is a highly efficient statistical technique for segmentation, or tree growing, developed by Kass (1980). The analysis in CHAID begins by dividing the population into two or more groups based on the categories of the "best" predictor of a dependent variable. It merges values that are judged to be statistically homogeneous (similar) with respect to the target variable, and maintains all other values that are heterogeneous (dissimilar). Each of these groups is then divided into smaller subgroups based on the best available predictor at each level.

The splitting process continues recursively until no more statistically significant predictors can be found (or until some other stopping rule is met).

The CHAID software displays the final subgroups (segments) in the form of a tree diagram whose branches (nodes) correspond to the groups. The segments that CHAID derives are mutually exclusive and exhaustive. It also produces a file of associated pseudocode that can be used in SAS®, with minor modifications, to create a SAS variable for indicating the groups (i.e., the nonresponse adjustment cells). A node will *not* be split if any of the following conditions are met:

- All cases in a node have identical values for all predictors.
- The node becomes pure; that is, all cases in the node have the same value of the target (or dependent) variable.
- The depth of the tree has reached its prespecified maximum value.
- The number of cases constituting the node are less than a prespecified minimum parent node size.
- The split at the node results in producing a child node whose number of cases is less than a prespecified minimum child node size.
- No more statistically significant splits can be found at the specified level of significance. It should be noted that all but the first two of these rules could be user specified. CHAID is not binary; that is, it can produce more than two categories at any particular level in the tree. Therefore, it tends to create a wider tree than do the binary growing methods.

Classification and regression tree builders (CART, cTree and rpart), as popularized by Breiman et al. (1984), make binary splits based on either yes/no or less than/greater than criteria. The choice of variable to be used at each split and the decision criteria are determined by 10-fold cross-validation. The resultant tree can be pruned. Results can be improved by making use of relative losses and a priori frequencies of occurrence.

The approach used in CART and cTree has two drawbacks:

- *Computational complexity.* An ordered variable with n distinct values at a node induces $(n - 1)$ binary splits. For categorical variables, the order of computations increases exponentially with the number of categories; it is $(2^{M-1} - 1)$ for a variable with M values.
- *Bias in variable selection.* Unrestrained search tends to select variables that have more splits. As Quinlan and Cameron-Jones (1995, 13:287–312) observe, "For any collection of training data, there are 'fluke' theories that fit the data well but have low predictive accuracy. When a very large number of hypotheses is explored, the probability of encountering such a fluke increases."

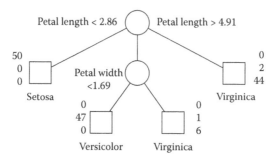

FIGURE 10.11
Iris data using the FACT method. The triple beside each terminal node gives the number of cases in the nodes of Setosa, Versicolor, and Virginica, respectively.

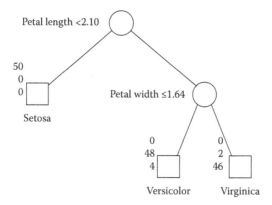

FIGURE 10.12
Iris data using the QUEST method. The triple beside each terminal node gives the number of cases in the nodes of Setosa, Versicolor, and Virginica, respectively.

One alternative is the FACT algorithm due to Loh and Vanichsetakul (1988). Instead of combining the problem of variable selection with that of split criteria, FACT deals with them separately. At each node, an analysis of variance F-statistic is calculated for each ordered variable. The variable with the largest F-statistic is selected, and linear discriminant analysis is applied to it to find the cutoff value for the split. Categorical variables are handled less efficiently by transforming them into ordered variables. If there are J classes among the data in a node, this method splits the node into J subnodes.

The QUEST (for Quick, Unbiased, Efficient, Statistical Tree) introduced by Loh and Shih (1997), like CART, yields binary splits and includes pruning as an option, but it has negligible variable selection bias, and retains the computational simplicity of FACT. QUEST can easily handle categorical predictor variables with many categories. It uses imputation instead of surrogate splits to deal with missing values. If there are no missing values in the data, QUEST can optionally use the CART algorithm to produce a tree with univariate splits.

Warning: While CART, FACT, and QUEST yield different error rates for the IRIS data, this may not be true for all data sets.

CRUISE, another enhanced version of FACT, introduced by Kim and Loh (2001), stands for classification rule with unbiased interaction selection and estimation. Like QUEST, it has negligible bias in variable selection and has several ways to deal with missing values.

Some Comparisons

Trujillano et al. (2009) used decision trees to classify the severity of critically ill patients. The decision tree they generated based on the CHAID methodology is shown in Figure 10.13. It uses seven variables and begins with the variable INOT. Fifteen decision rules were generated with an assignment rank of probability ranging from 0.7% to a maximum of 86.4%. The Glasgow value, age, and (A-a)O_2 variables were divided into intervals with more than two possibilities.

The decision tree generated by Trujillano et al. (2009) shown in Figure 10.14 based on the CART methodology uses only five variables and also begins with INOT. It generates eight decision rules with an assignment rank of probability ranging from 5.9% to a maximum of 71.3%.

Their C4.5 model depicted in Figure 10.15 uses 6 variables (the 5 common variables and mean arterial pressure) and generates 10 decision rules. The probabilities ranged between 7.6 and 76.2%. In contrast to the other decision trees, the first variable was the point value on the Glasgow scale.

Attempting to minimize the overall length of the decision tree was found by Comley and Dow (2003) to produce more errors in classification than C5.

Reducing the Rate of Misclassification

In this section we discuss the use of boosting and ensemble methods to reduce errors in misclassification.

Boosting

C5, a much enhanced version of C4.5, provides the ability to improve on the original decision tree via boosting. Instead of drawing a succession of independent bootstrap samples from the original instances, boosting maintains a weight for each instance—the higher the weight, the more the instance influences the next classifier generation, as first suggested by Quinlan (1996).

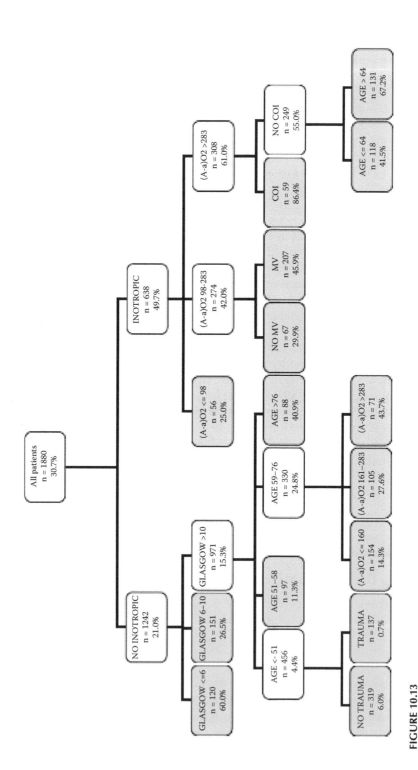

FIGURE 10.13

Classification tree by CHAID algorithm. The gray squares denote terminal prognostic subgroups. INOT, inotropic therapy; (A-a)O2 gradient, alveolar-arterial oxygen gradient (mmHg); MV, mechanical ventilation; COI, chronic organ insufficiency.

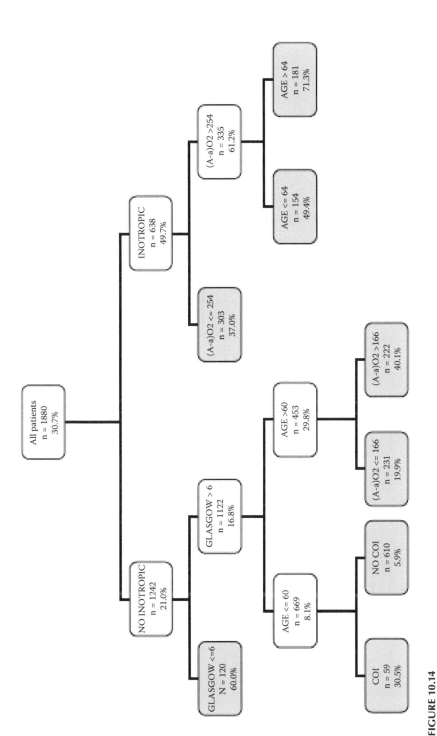

FIGURE 10.14

Classification tree by CART algorithm. The gray squares denote terminal prognostic subgroups. INOT, inotropic therapy; (A-a)O2 gradient, alveolar-arterial oxygen gradient (mmHg); MV, mechanical ventilation; COI, chronic organ insufficiency.

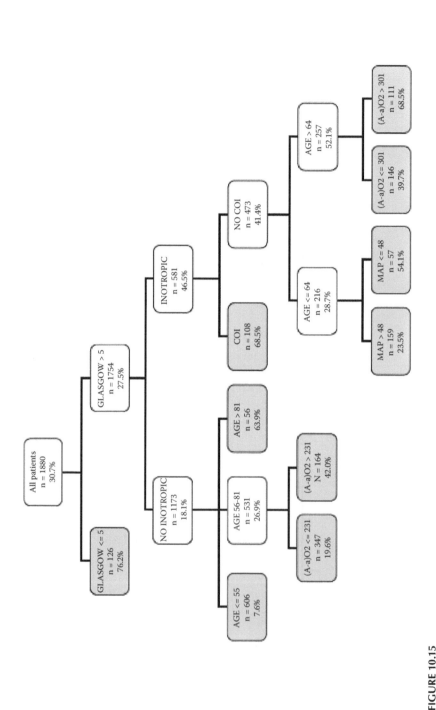

FIGURE 10.15

Classification tree by C4.5 algorithm. The gray squares denote terminal prognostic subgroups. INOT, inotropic therapy; (A-a)O2 gradient, alveolar-arterial oxygen gradient (mmHg); MV, mechanical ventilation; COI, chronic organ insufficiency; MAP, mean arterial pressure.

Initially, all weights are set equally, but at each trial, the vector of weights is adjusted to reflect the performance of the corresponding classifier, with the result that the weight of misclassified instances is increased so that the weak learner is forced to focus on the hard instances in the training set.

AdaBoost Algorithm

1. Initialize some weights for individual sample points: $w_i^0 = 1/n$ for $i = 1, \ldots, n$. Set $m = 0$.

2. Increase m by 1. Fit the base procedure to the weighted data; that is, do a weighted fitting using the weights w_i^{m-1} yielding the classifier $\hat{g}^m(.)$.

3. Compute the weighted in-sample misclassification rate

$$err^m = \sum_{i=1}^{n} w_i^{m-1} I\{Y_i \neq \hat{g}^m(X_i)\} / \sum_{i=1}^{n} w_i^{m-1}$$

$$\alpha^m = \log(1 - err^m) - \log(err^m)$$

and update the weights

$$\bar{w}_i = w_i^{m-1} \exp[\alpha^m I\{Y_i \neq \hat{g}^m(X_i)\}]$$

$$w_i^m = \bar{w}_i \Big/ \sum_{j=1}^{n} \bar{w}_j$$

4. Iterate steps 2 and 3 until $m = m$stop, and build the aggregated classifier by weighted majority voting:

$$\hat{f}_{AdaBoost}(x) = \arg\max_{y \in [0,1]} \sum_{m=1}^{m_{stop}} \alpha^m I\{\hat{g}^m(x) = y\}$$

Ensemble Methods

Ensemble methods combine multiple classifiers (models) built on a set of resampled training data sets, or generated from various classification methods on a training data set. This set of classifiers forms a decision committee, which classifies future samples. The classification of the committee can be by simple vote or by weighted vote of individual classifiers in the committee.

The essence of ensemble methods is to create diversified classifiers in the decision committee. Aggregating decisions from diversified classifiers is an effective way to reduce bias existing in individual trees. However, if classifiers in the committee are not unique, the committee has to be very large to create diversity within it.

Comparison of Classification Tree Algorithms

Feature	CRUISE	QUEST	tree()	CART	CHAID	C4.5	C5
Split Variable Selection							
Unbiased selection	y	y					
Pairwise interaction detection	y						
Split Types							
Univariate (axis-orthogonal)	y	y	y	y	y	y	y
Linear combinations (oblique)	y	y					
Gain ratio splitting						y	y
Gini reduction or twoing	y	y	y	y			
Miscellaneous							
Choice of misclassification costs	y	y		y	y		y
Weight cases							y
Choice of class prior probabilities	y	y		y			
Choice of impurity functions		y		y		y	y
Bivariate discriminant node models	y						
Bagging				y			
Error estimation by cross-validation	y	y	y	y	y		
Boosting							y
Eliminate barely relevant attributes							y
Number of Branches at Each Node							
Always two		y	y	y			
Two or more	y				y	y	y
Missing Value Methods							
Imputation	y	y					
Alternate/surrogate splits	y		y	y			
Missing value branch					y		
Probability weights						y	y
Tree Size Control							
Stopping rule					y		

Feature	CRUISE	QUEST	tree()	CART	CHAID	C4.5	C5
Prepruning						y	y
Test sample pruning	y	y		y			y
Cross-validation pruning	y	y	y	y			y
Tree Diagram Formats							
Text	y	y	y	y		y	y
LATEX	y	y					
allClear	y	y					
Proprietary				y	y		
Platforms							
Windows	y	y	y	y	y	y	y
Macintosh		y					
Linux	y	y	y	y		y	y
Sun	y	y	y	y		y	y
Free compiled code	y	y					
Free source code			y			y	

Software for Creating Decision Trees

CHAID is part of IBM SPSS AnswerTree. See http://www.spss.com/software/statistics/answertree/.

CRUISE may be downloaded without charge from http://www.stat.wisc.edu/~loh/cruise.html.

QUEST may be downloaded without charge from http://www.stat.wisc.edu/~loh/quest.html.

tree(), an R program, may be downloaded from within R by typing install.packages ("tree"); recommended are the more comprehensive tree building routines available by downloading the rpart package from http://cran.r-project.org/web/packages/rpart/index.html.

A trial version of CART may be obtained from http://salford-systems.com.

C4.5 may be downloaded without charge from http://www2.cs.uregina.ca/~hamilton/courses/831/notes/ml/dtrees/c4.5/tutorial.html.

A scaled-down version of See5 may be downloaded without charge from http://www.rulequest.com/download.html. Its predictions are purely categorical. Cubist, which may be downloaded from the same site, yields continuous, regression-like predictions.

SAS PROC TREE has a misleading name, as it does not yield a decision tree, but makes use of the nearest-neighbor procedure described by Wong and Lane (1983) to form clusters.

Excel users can form decision trees by downloading Ctree, a macro-filled Excel spreadsheet, from http://www.geocities.com/adotsaha/CTree/CtreeinExcel.html.

Validation vs. Cross-Validation

Goodness of fit is not prediction. While cross-validation is invaluable for the development of a classification scheme, the true validation of such a scheme comes only when it is able to correctly classify a set of observations that is completely distinct from the set that was used to develop the classification scheme. Absent such validation, any published comparison of different methods must be regarded with great suspicion.

Oehler et al. (2010) divided their data into two parts. They drew a random sample (without resampling) of one-eighth of the observations from the main data set and termed this subsample the test set; the remaining data became the training set.

They applied three separate classification methods to the training set and developed three independent classifiers using 10-fold cross-validation. They validated the result by applying the three classifiers to the test set and recorded the results.

They repeated this procedure 33 times, that is, the random division into training and test set was repeated 33 times, and prepared a box plot of the results as shown in Figure 10.16. The box plot labeled BRT5 is that of a boosted decision tree. It also may be compared with the other classification methods—artificial neural network (ANN) and NEMIS (a nonlinear function) based on its median result (tied for best), its consistency (best), or its worst case (second worst of the three methods).

An improved method of validation using restricted permutations is considered in the "Model Validation" section in Chapter 11.

Summary

Decision trees offer the advantage over regression methods in that their output lends itself to cause-and-effect interpretation. They should be used whenever the following is true:

- Predictors are interdependent and their interaction often leads to reinforcing synergistic effects.
- A mixture of continuous and categorical variables, highly skewed data, and large numbers of missing observations adds to the complexity of the analysis.

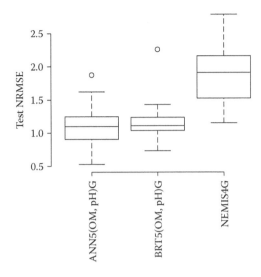

FIGURE 10.16
Box plots of results reported by Oehler et al. (2010)

Improved efficiency may be brought about through the use of ensemble methods. Many competing algorithms exist. While cross-validation may be used for the development of decision trees, comparison of the various algorithms or of decision trees with regression methods requires true validation through the use of a separate distinct test set(s).

To Learn More

Automated construction of a decision tree by repeated resampling dates back to Morgan and Sonquist (1963). The CHAID algorithm was introduced by Kass (1980). Breiman et al. (1984) popularized CART. Loh and Shih (1997) introduced the QUEST algorithm. Kim and Loh (2001) is the principal reference for the CRUISE algorithm. Clark and Pregibon (1992) provide a probability basis for decision trees. Comparisons of the regression and tree approaches were made by Nurminen (2003) and Perlich et al. (2003). Lim et al. (2000) compare the prediction accuracy of 33 decision tree algorithms. A primer on decision trees may be seen in Kingsford and Salzberg (2008). For an in-depth discussion of boosting algorithms, see Bühlmann and Hothorn (2007). Arditi and Pulket (2004) use boosted decision trees to predict the outcome of construction litigation.

Exercises

1. Fit a decision tree to data you've already modeled via stepwise regression. Does it use the same variables? Investigate the effects of pruning.

2. Using the data from http://www.sgi.com/tech/mlc/db/breast.test, build a decision tree to forecast malignancy. A description of this database is located at http://www.sgi.com/tech/mlc/db/breastLoss.names.

3. Analyze the data located at http://www.cba.ua.edu/~mhardin/hepatitisdatasets/ by both stepwise regression and decision tree methods. Validate your findings by one or more methods.

4. Analyze one or more of the data sets you find at http://archive.ics.uci.edu/ml/.

11

Restricted Permutations

In previous chapters, it became obvious that applying permutation methods is not quite so straightforward a process as was first described in Chapter 4. We need to establish *both* the set of labels that are to be arranged and the test statistic to be employed. As we saw in the "Crossover Designs" and "Which Sets of Labels Should We Rearrange?" sections in Chapter 6, the set of labels to be arranged will depend upon the hypothesis to be tested.

In this chapter, we consider some recent developments in permutation theory where the number of permutations is restricted to a subgroup of the possible rearrangements. The results are tests for quasi-independence, complete factorials, interactions in multifactor designs, and classification accuracy.

Quasi Independence

Elderton et al. (1913) studied the incidence of birth defects in English families. They addressed the question of whether there was a correlation between an imbecile's birth order and the size of his family. In these authors words, "Clearly the size of the family must always be as great or greater than the imbecile's place in it, and the correlation table is accordingly one cut off at the diagonal and there would certainly be correlation, if we proceeded to find it by the usual product moment method, but such correlation is or clearly may be wholly spurious." Thus, the procedures of Chapter 7 won't help us, as all the entries in the upper off-diagonal portion of the table are zero.

We can apply the following algorithm provided by Diaconis et al. (2001, 195–222) to generate a random rearrangement of the triangular $R \times C$ table with entries T_{ij}, row sums $r_1, \ldots, r_R$, and columns sums $c_1, \ldots, c_C$:

- Place c_1 balls labeled 1 in an urn and sample r_1 of these without replacement. Set T^*_1 equal to the number of balls in the sample labeled 1.
- Add c_2 balls labeled 2 to the urn. Sample r_2 of these without replacement. Set T^*_{2k} equal to the number of balls in the sample labeled k in the sample, where $k = {}_{1 \text{ or }} 2$.
- Continue this process until all c_C balls labeled C have been added to the urn.

We repeat this procedure several hundred times, computing a summary statistic for each table and comparing it with the value of the statistic for the original table.

Note that very similar upper triangular tables arise in genetics in testing goodness of fit of the Hardy-Weinberg equilibrium model.

Chen and Liu (2007) make use of importance sampling in generating permutations with restricted positions for testing for quasi-independence in more complex situations.

Complete Factorials

In unreplicated factorials, we have no degrees of freedom with which to estimate the residual sum of squares from the usual analysis of variance by linear models, so that the parametric F-test is not applicable. In this section, we consider a resampling alternative due to Valle et al. (2002).

Define the *design matrix* $\mathbf{XN}$, $N = 2k$, of a $2k$ complete factorial design as

$$\mathbf{XN} = [\mathbf{1}\ \mathbf{D}\ \mathbf{PD}]$$

where $\mathbf{1}$ is the column vector of the general mean m, which is composed of all +1 elements, $\mathbf{D}$ is a $N \times k$ matrix whose rows contain all combinations of two distinct levels (±1) of the k main factors, and $\mathbf{PD}$ is a $N \times [N - k - 1]$ matrix whose columns correspond to the t-factor interactions, and these columns are Kronecker products of columns corresponding to t main factors, $2 \le t \le k$. The usual linear model for the analysis of variance of a $2k$ complete factorial design with fixed effects is $\mathbf{y} = \mathbf{XN}\boldsymbol{\beta} + \mathbf{e}$, where $\mathbf{XN}$ is the square design matrix, $\mathbf{y}$ is an $N \times 1$ vector of responses, $\boldsymbol{\beta}$ is an $N \times 1$ vector of parameters, and $\mathbf{e}$ is an $N \times 1$ vector.

For example,

$$X_4 = [\mu, A, B, AB] = \begin{bmatrix} + & + & + & + \\ + & + & - & - \\ + & - & + & - \\ - & - & - & + \end{bmatrix}$$

Loughin and Noble (1997) showed that by permuting the whole vector of observations, the test on the largest absolute effect (identified from the least-squares estimates) is the *only* exact one. Applying the restricted permutation strategy of Pesarin and Salmaso (2010) permits us to perform exact permutation tests for a larger number of effects. The hypotheses to test are $H_0\beta_1$:

$\{\beta_1 = 0\}$ against $H_1\beta_1$: $\{\beta_1 \neq 0\}$, irrespective of whether or not $H_0\beta_2 \cup H_0\beta_3 \cup \ldots \cup H_0\beta_{N-1}$ are true, and so forth.

Pesarin and Salmaso (2010) propose the following procedure:

1. Write the design matrix in normal form.
2. Exchange rows in the design matrix such that in the second row columns corresponding to k factors have the same configuration of ± 1's as the rth row of the realigning matrix, $1 \leq r \leq N - 1$.
3. Exchange rows in the design matrix such that each pair of elements e_1, e_2 in two adjacent rows of each of the $k - t$ factors with +1 in the second row has the same sign, that is, either $e_1 = e_2 = +1$ or $e_1 = e_2 = -1$.

Each realignment identifies 2^{k-1} nonaligned factors. Each pair of distinct realignments identifies 2^{k-2} nonaligned factors. For each realignment, we can perform a permutation test on the largest effect by considering as a test statistic the absolute value of the estimate of the effect.

Unfortunately, the number of possible permutations is limited. For example, with four factors, there are only 16 possible permutations, and the minimum possible significance level is 1/16 > 5%. To perform the test for larger designs, first download the rcode for t2p.r, create_design.r, and unreplicated.r from http://www.gest.unipd.it/~salmaso/web/springerbook.htm.

```
Set X = create_design (k);
B = rep(0, length(data))
y = X%*%b + data
Note length(data) must be 2**k.
Run unreplicated (y, X,,IER=0.05,,B=1000)
```

where IER is the individual error rate and B is the number of rearrangements considered.

Synchronized Permutations

Let's take a closer look at the confounding between main effects and interactions that we considered in the "Testing for Interactions" section of Chapter 6.

Figure 11.1a depicts a two-factor experimental design with two levels of one factor and three of the second. Within each row and column, the interaction terms sum to zero because of the way we have defined them as deviations from the sum of the main effects. If we were to compare the expected values of the row sums, the difference would depend solely on the row effects, if any. Similarly, any differences in the expected values of the column sums would depend solely on the column effects, if any.

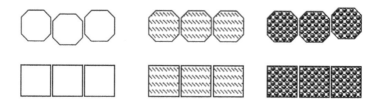

FIGURE 11.1a
A 2×3 design with three observations per cell.

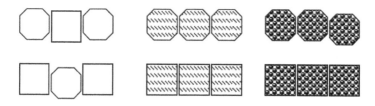

FIGURE 11.1b
A 2×3 design with three observations per cell.

The situation is quite different in Figure 11.1b, where row and row-column interactions are clearly confounded.

In Figure 11.2.a the row sum of the interaction terms is still zero, so that once more any differences in the expected values of the row sums would depend solely on the row effects, if any. Similar remarks involving columns rather than rows apply to Figure 11.2b. By restricting ourselves to similar synchronized rearrangements, we can avoid confounding interactions with main effects.

Note that the original set of all possible rearrangements

$$\binom{4m}{m \ \ m \ \ m}$$

in number is now subdivided into a set of synchronized rearrangements,

$$\sum_{k=0}^{m} \binom{m}{k}^4$$

in number, for testing row effects, two sets of similar size for testing column effects and interactions, respectively, and the unsynchronized permutations. Only the original and not yet permuted observations of Figure 11.1a are common to more than one set of synchronized observations, so that by restricting ourselves to one of these sets each time, we obtain independent exact tests for row effects, column effects, and interactions.

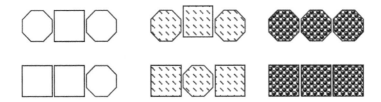

FIGURE 11.2a
A 2 × 3 design with three observations per cell after a synchronized permutation between rows.

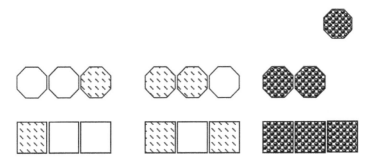

FIGURE 11.2b
A 2 × 3 design with three observations per cell after a synchronized permutation between columns.

Generalizing These Results to Multiple Factors

Extending the definition of a two-factor experimental design given in Good (2002), we define a M-factor *experimental design* as an M-dimensional point lattice L such that with each point of the lattice $\mathbf{s} = (i_1, i_2, \dots i_M)$, $i_1 = 1, \dots, I_1$; $i_m = 1, \dots, I_M$ is associated with a set of independent random variables $X_k(\mathbf{s})$; and $k = 1, \dots, n\mathbf{s}$ identically distributed as $F(x - \Delta\mathbf{s})$, where

$$\Delta\mathbf{s} = E\, X_k(\mathbf{s}) = \mu + \sum_{m=1}^{M} \alpha_{i_m} + \sum_{m-1}^{M} \sum_{j=1}^{m-1} \beta_{i_j i_m} + \dots \qquad (8.1)$$

$$\sum_{i_m=1}^{I_m} \alpha_{i_m} = 0 \ \text{ for all } m \qquad (8.2)$$

$$\sum_{i_j=1}^{I_j} \beta_{i_j i_m} = \sum_{i_m}^{I_m} \beta_{i_j i_m} = 0 \ \text{ for all } m \text{ and } j \qquad (8.3)$$

and so forth.

If $n\mathbf{s} = n$ for all $\mathbf{s}$, we term the design *balanced*.

A well-known result from group theory is that every permutation can be constructed from a succession of pairwise exchanges. Thus, one way to convert $\{A, B, C\}$ to $\{B, C, A\}$ is to apply the changes $(1, 3)(1, 2)$ in that order. The pairwise exchanges we will be concerned with are those rearrangements in which an observation in one cell of an experimental design is swapped with an observation in another cell at the same location in the design, with the exception of the lth coordinate. For example, in a two-factor design, such an exchange might be between two cells in the same row but different columns. We let $(l; l1, l2; k_1, k_2; s)$ denote a pairwise exchange in which we swap the k_1th observation at $(i_1, \ldots, i_{l-1}, l1, \ldots, i_M)$ with the k_2th observation at $(i_1, \ldots i_{l-1}, l2, \ldots, i_M)$.

We let $(t: l; l1, l2; s)$ denote an exchange in which t such pairwise exchanges take place simultaneously between the same pair of cells. In a balanced design, for each fixed pair of cells $(l; l1, l2)$ there are exactly $\binom{n}{t}^2$ such exchanges.

Let r_l denote the vector $(i_1, \ldots, i_{l-1}, i_{l+1}, \ldots, i_M)$ derived from s by eliminating the lth coordinate and reorder the coordinates of s so that $s= \{l, r_l\}$. A synchronized pairwise rearrangement used for testing main effects, for example,

$$H_l : \alpha_{l_j} = 0 \text{ for } l_j = 1, \ldots, I_l$$

has the form

$$(t: l; l1, l2) = \Pi r_l \, (t: l; l1, l2; (l, r_l))$$

That is, any exchanges of observations between levels $l1$ and $l2$ of the lth factor take place at the same time at *all* levels of all the remaining factors.

A synchronized pairwise rearrangement used for testing the interaction between factors l and j has the form $(t: l; l1, l2) \, (t: j; j1, j2)$, in which the initial simultaneous pairwise exchanges made between rows $j1$ and $j2$ at each combination of all other factors in the design are followed by simultaneous pairwise exchanges between columns $l1$ and $l2$.

Synchronized rearrangements are composed of combinations of synchronized pairwise rearrangements involving distinct pairs of rows and columns.

By restricting attention to synchronized rearrangements in the two-factor case, Salamaso (2003) and Pesarin (2001) are able to define three subsets of rearrangements P_R, P_C, and P_{RC} such that the only point they have in common is the identity transformation I,

$$P_{RC} \subsetneq P_R = P_{RC} \subsetneq P_C = P_R \subsetneq P_C = I \tag{8.4}$$

P_R, for example, is the set of all synchronized rearrangements among rows only.

Let l denote the factor corresponding to rows; P_R would include all synchronized pairwise exchanges such as $(t: l; l1, l2)$ for $0 \le t \le \min(n_{l1}, n_{l2})$, along with products (in the permutation sense) of any number of such synchronized pairwise exchanges.

Condition 4 implies that a test based upon a permutation distribution generated by the elements of P_R will be completely independent of tests generated by the elements of P_C or P_{RC}. Through weak exchangeability with respect to P_R, P_C, and P_{RC}, Pesarin (2001) and Good (2002) are able to derive separate uncorrelated tests of the null hypotheses concerning the two main effects and the interactions, albeit for completely different test statistics.

Algorithms

To generate random rearrangements for use in testing for a row main effect, theorem 1 of Appendix B shows us that we may proceed as follows:

1. Compute $m_j = \min_i n_{ij}$ for all columns j, where n_{ij} is the number of observations in the ith row and jth column.
2. Set $V = c(\text{rep}(1, m_1), \text{rep}(2, m_2), \ldots, \text{rep}(J, m_J))$, where $\text{rep}(1, 3) = 1,1,1$.
3. Permute the elements of V to form πV.
4. Permute the contents of each cell in the design within the cell. (This step ensures that each observation in a cell will have an equal probability of being relabeled during step 5.)
5. Proceed a row at a time, creating a new data matrix as follows: Remove the first m_i elements of each cell in the row and place in a temporary vector. Rearrange the elements in the vector as specified by the indices in πV. Place the rearranged elements back in the row.

In similar fashion, we can generate any number of random rearrangements to use in testing for a column effect. The only difference is that we need to first compute $_im = \min_j n_{ij}$

To generate random rearrangements for use in testing for a row-column interaction, theorem 2 of Appendix B tells us that we need to proceed in three stages:

At the first stage, we generate one of the random rearrangements used in testing for a column effect. The resultant relabeled design matrix can be written as the sum of two matrices, one containing all the relabeled elements, and one containing all the elements whose labels have remained unchanged.

At the second stage, we repeat steps 1–5 independently for each of the two matrices.

At the third stage, we compute the value of the test statistic for the matrix formed by summing the two rearranged matrices.

Which Test Should We Use?

The analysis of variance (ANOVA) applied to $R \times C$ experimental designs yields almost exact tests regardless of the distribution from which the data are drawn (Jagers, 1980). Of course, because the ANOVA tests for main effects and interaction share a common denominator—the within-sum of squares—the resultant p-values are positively correlated. Thus, a real non-zero main effect may be obscured by the presence of a spuriously significant interaction.

Although tests based on synchronized permutations are both exact and independent of one another, there are so few synchronized permutations in the two-factor design with small samples that these tests lack power. For example, in a 2×2 design with three observations per cell, there are only nine distinct values of each of the test statistics.

Model Validation

Ojala and Garriga (2010) considered the possibility that one of the algorithms described in the previous two chapters may have identified some pattern in the high-dimensional data of biomedical and satellite images that resulted from chance alone. They contrasted two methods of validation using permutation tests. In the first traditional method, the class labels on the observations are randomly permuted to see if this will affect the predictions. This method almost always validates the classification, even in instances when the distinction between classes is weak or spurious.

In their second method, they applied independent permutations to the predictors within each class as follows:

Let X be an $m \times n$ data matrix in which each row is a sample and each column corresponds to a different variable. Let Y be the corresponding $m \times 1$ matrix of the known classifications.

Let $X(c)$ be the submatrix of X with class label c, that is, $X(c) = \{X_i | y_i = c\}$ of size $l_c \times m$. Let $X^j(c)$ denote the jth column of this matrix corresponding to the jth predictor.

Let $\pi_{c1}, \ldots, \pi_{cm}$ be m independent permutations of the numbers $\{1, \ldots, l_c\}$.

Let $\pi_c X(c)$ be a randomized version of $X(c)$, where each π_{cj} is applied independently to the column vector $X^j(c)$. That is, $\pi_c X(c) = \{\pi_{c1} X^1(c), \ldots, \pi_{cm}(X_m(c))\}$.

If the variables that compose X are interdependent, the effect of $\pi = \{\pi_1 \ldots \pi_C\}$ on X is to destroy this interdependence. If the classifier we employed exploits the interdependency between the features in the data, then the training error for the permuted data will be larger than the training error for the original data. To validate our model, we derive the permutation distribution of the training error by performing a series of random permutations as described above. We use this distribution to establish whether the classifier uses the possibly existing feature dependency to improve the classification accuracy.

Exercises

1. Are the main effects and interactions significant in the following table taken from Hettmansperger (1984)? Survival times are measured for each of two different poisons and two different treatments, and the experiment is replicated four different times.

Survival Times Following Treatment		
	T1	T2
P1	31, 45, 46, 43	45, 71, 66, 62
P2	22, 21,18, 23	30, 36, 31, 33

Appendix A: Basic Concepts in Statistics

Additive vs. Multiplicative Models

In an additive model, the contributions of the various factors add up to the final result; in a multiplicative model, changes in the contributing factors result in percentage changes in the final result. We can convert a multiplicative model to an additive one by taking logarithms. Unfortunately, if the errors had a normal distribution originally, they will have a lognormal distribution after taking logarithms. Fortunately, the resampling methods are not affected by the change, as they are distribution-free.

Central Values

Some would say that the center of a distribution is its 50th percentile or *median*, that value P_{50} such that half the population has equal or smaller values and half the population has greater values. Others, particularly physicists, would say that it is the *arithmetic mean* $\bar{X}$ or balance point, such that the sum of the deviations about that value is 0. In symbols, we would write $\sum(X-\bar{X})=0$. A second useful property of the arithmetic mean is that the sum of the squared deviations about any value K, $\sum(X-K)^2$, is a minimum when we chose $K = \bar{X}$.

Which parameter should you use? We'd suggest the median, and here are two reasons why:

1. When a cumulative distribution function is symmetric as in the figure, the arithmetic mean and the median have the same value. But even then, one or two outlying values s, typographical errors, for example, would radically affect the value of the mean, while leaving the value of the median unaffected. For example, compare the means and medians of the following two sets of observations: 1, 2, 4, 5, 6, 7, 9 and 1, 2, 4, 5, 6, 7, 19.

2. When a cumulative distribution function is highly skewed, as with incomes and house prices, the arithmetic mean can be completely misleading. A recent *LA Times* article featured a great house in

Beverly Park at US$80 million. A single luxury house like that has a large effect on the mean price of homes in its area. The median house price is far more representative than the mean, even in Beverly Hills.

The *geometric mean* is the appropriate choice of central value when we are expressing changes in percentages, rather than absolute values. For example, if in successive months the cost of living was 110, 105, 110, 115, 118, 120, and 115% of the value in the base month, the geometric mean would be $(1.1*1.05*1.1*1.15*1.18*1.2*1.15)^{1/7}$.

Use of the geometric mean is also recommended when tracking populations of viruses and bacteria, populations whose growth, if unchecked, can vary in size by several orders of magnitude.

Combinations and Rearrangements

We can choose three people out of six to receive a treatment in 20 ways. Here is why: we can choose the first person in six ways, the second person in five, and the third person in four. In symbols, this would be the same as 6!/3!. But notice that choosing Bill first, then Andy, and then Sarah would lead to exactly the same result as if we'd first chosen Andy, then Bill, and then Sarah. In fact, all 3 × 2 orderings would have led to the same sample. So the real number of arrangements or combinations is 6!/(3!3!).

Dispersion

The mean or median only tells part of the story. Measure my height using a tape measure three times and you'll probably come up with readings of 69, 70, and 71 in. Measure my diastolic blood pressure three times and you might come up with readings as diverse as 80, 85, and 90 mm. Clearly, the latter set of readings is far more dispersed, and thus less precise, than the first.

One measure of dispersion is the population *variance*, the sum of the squared deviations about the population mean, $\sum_{i=1}^{n} (X_i - \bar{X})^2 / (n-1)$. Unfortunately, the units in which the variance is defined are the squares of the units in which the original observations were recorded. In consequence, we are more likely to use the *standard deviation*, defined as the square root of the variance, as it has the same units as the original measurements. The ratio of the standard deviation to the mean (known as the *coefficient of variation*) is dimensionless and independent of the units in which the original observations were recorded.

If the frequency distribution we are studying has the *normal distribution* depicted in the figure, then the probability that an observation will lie within one 1.64 standard deviations of the mean is 90%. Observations that are made up of the sum of a large number of factors, each of which makes only a very small contribution to the total, will always be normally distributed.

If the distribution is not normal, nor has some other well-tabulated standard form, then knowledge of the standard deviation is far less informative. The *interquartile range* tells us the length of the interval between the 25th and the 75th percentile. In other cases, we might want to know the values of the 25th and 75th percentiles or of the 10th and 90th. Figure 2.2 depicts a *box and whiskers plot* of 22 observations in which we can see at a glance the values of the 25th and 75th percentiles (the box), the median (the bar inside the box), and the *minimum* and *maximum* (the ends of the whiskers).

Some statisticians also favor the use of the L_1 norm, defined as

$$\sum_{i=1}^{n} |X_i - \bar{X}| / (n-1)$$

Frequency Distribution and Percentiles

The *cumulative distribution function* $F[x]$ is the probability that an observation drawn at random from a population will be less than or equal to x. The kth percentile of a population is that value P_k such that $F[P_k] = k$ percent. There are many economic, health, and political applications where our primary interest is in these *percentiles*, or *quantiles* as they are sometimes called.

Linear vs. Nonlinear Regression

Linear regression is linear in its coefficients and takes the linear form

$$EY = \mu + \beta_1 f_1[X_1] + \beta_2 f_2[X_2] + \dots$$

where EY is the expected value of Y for the given values of the predictors, and $f_1, f_2, \dots$ are specific functions of the predictors, for example, $f_1[X_1] = X_1$, $f_2[X_2] = \log[X_2]$ and so forth.

$EY = \mu + \log [\beta X]$ is an example of nonlinear regression. There is a unique optimal solution for a linear regression equation; to find the optimal solution for a nonlinear equation requires a series of iterations, and the final result may depend on the initial guess.

Regression Methods

With ordinary least squares regression, our model is that $EY = a + bX$, and we seek to minimize the sum

$$\sum_i \left(Y_i - a - bX_i\right)^2$$

With least absolute deviation regression, our model is that $EY = a + bX$, and we seek to minimize the sum

$$\sum_i \left|Y_i - a - bX_i\right|$$

In estimating the τth quantile via quantile regression, we try to find that value of β for which $\Sigma_{k\rho\tau}(y_k - f[x_k, \beta])$ is a minimum, where

$$\rho\tau[x] = \tau x \text{ if } x > 0$$

$$= (\tau - 1)x \text{ if } x \leq 0$$

For the 50th percentile or median, the result is equivalent to that obtained via LAD regression.

Appendix B: Proof of Theorems

As in Chapter 8, we define an M-factor *experimental design* as an M-dimensional point lattice L such that with each point of the lattice $\mathbf{s} = (i_1, i_2, ..., i_M)$, $i_1 = 1, ..., I_1$; $i_m = 1, ..., I_M$ is associated with a set of independent random variables $Xk(\mathbf{s})$; and $k = 1, ..., ns$ is identically distributed as $F(x-\Delta\mathbf{s})$, where

$$\Delta\mathbf{s} = E\,X_k(\mathbf{s}) = \mu + \sum_{m=1}^{M} \alpha_{i_m} + \sum_{m=1}^{M} \sum_{j=1}^{m-1} \beta_{i_j i_m} + ... \tag{B.1}$$

$$\sum_{i_m=1}^{I_m} \alpha_{i_m} = 0 \ \text{ for all } m \tag{B.2}$$

$$\sum_{i_j=1}^{I_j} \beta_{i_j i_m} = \sum_{i_m=1}^{I_m} \beta_{i_j i_m} = 0 \ \text{ for all } m \text{ and } j \tag{B.3}$$

and so forth.

Let $\mathbf{X}$ be the vector space formed from finite combinations of the random variables $\{X_j(\mathbf{s})\}$ that form an experimental design. We define a semimetric ρ on $\mathbf{X}$ such that if $X, Y \, \varepsilon \, \mathbf{X}$ with distribution functions F_X, F_Y, then $\rho(X, Y) = sup_z |F_X(z) - F_Y(z)|$.

In line with group-theoretic convention, we term a rearrangement π a *similarity* if for all points W, X, Y, and Z in $\mathbf{X}$, $\rho(X,Y) = \rho(W,Z)$ if and only if $\rho(\pi X, \pi Y) = \rho(\pi W, \pi Z)$.

In what follows, dot notation will be used to represent a sum. Thus, $X_. = \Sigma X_i$ and $X_{.j.} = \Sigma_i \Sigma_k X_{ijk}$.

THEOREM 1

Let $\{X_j(\mathbf{s})\}$ be a set of independent observations in a complete balanced experimental design with factors in S such that $X_j(\mathbf{s}) = \Delta\mathbf{s} + \varepsilon_j(\mathbf{s})$, where $\Delta\mathbf{s}$ satisfies conditions in Equations B.1 to B.3 and the $\{\varepsilon_j(\mathbf{s})\}$ are a set of independent identically distributed random variables with mean zero. Then if π is a synchronized rearrangement with respect to the factor i, then π is a similarity under the null hypothesis $\alpha_{\iota_j} = 0$ for $\iota_j = 1,...,I_j$.

PROOF

Without loss of generality, suppose we designate as rows the factor for which we wish to test a main hypothesis, and columns a second factor that we wish

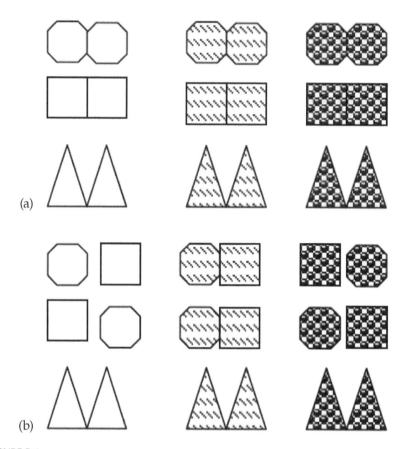

FIGURE B.1
(a) Part of an M-factor design. $M - 2$ of the dimensions are hidden in this representation. (b) The same design after an exchange of row elements. Similar synchronized exchanges have taken place in the hidden dimensions.

to test for a first-order interaction with rows. To simplify the notation, we assume that the factor subscripts are reordered so that the subscripts denoting rows and columns are in the initial positions.

Let π_{12} denote a pairwise synchronized exchange between rows 1 and 2 as in Figure B.1a and b. Of course, only two dimensions and three of the rows and columns are illustrated in our diagram. Similar synchronized exchanges have taken place at all levels of the remaining factors.

To generalize, suppose we make a pairwise synchronized exchange between rows m and p. If $i \neq m$ and $i \neq p$, then $E\pi_{mp}X_{i...} = EX_{i...}$ as the remaining factors in Equation B.1 vanish in accordance with the conditions in Equations B.2 and B.3. If $i = m$, then $E\pi_{mp}X_{i...} = EX_{i...} - I(\alpha_{Im} - \alpha_{Ip}) - \beta_{IJi}$. But if H_I is true, $\beta_{IJi} = 0$, so that $E\pi_{mp}X_{i...} = EX_{i...}$ Similarly for $i = p$ when H_1 is true. The $\varepsilon_j(\mathbf{s})$ are independent and identically distributed; the pairwise exchange involved the swap of equal numbers of independent identically distributed variables. So π_{mp} is a similarity. Since all synchronized exchanges are made up of similar pairwise synchronized exchanges, all are similarities, as was to be proved.

COROLLARY 1

The previous results extend immediately to an unbalanced experimental design providing there is at least one observation per cell.

The proof of the corollary is immediate, as the proof of the main theorem did not require that the design be balanced.

Imbalance has practical implications, as it will result in fewer opportunities for an exact test without randomization on the boundary and a reduction in power. Consider a 2×2 experimental design with 10 observations in three of the cells and 1 in the fourth. Only 1,001 synchronized rearrangements are available to test for a row effect—one for the identity plus

$$1^* \binom{10}{1}^3$$

A thousand distinct synchronized rearrangements plus the identity are available to test for a column effect and 1,000 for interaction. The total of 3,001 distinct synchronized rearrangements is a miniscule fraction of the

$$\binom{31}{10 \ 10 \ 10}$$

possible rearrangements for this design.

At least two observations per cell are essential if we are to be guaranteed independent tests of all factors.

Typically, the loss function associated with a testing problem will be symmetric about zero and monotone nondecreasing on the positive half line.

COROLLARY 2

Suppose g is a monotone nondecreasing function such as $g(i) = i$ or $g(i) = \log(i + 1)$. Then the distribution of the statistic $S_I = \sum_{i=1}^{I_i} g[i]X_{.i.}$ obtained from synchronized permutations can be used to obtain an exact, unbiased test of a hypothesis concerning a main effect in an M-factor design, such as $H_I: \alpha_I j = 0$ for all j against an ordered alternative, such as $K_I: \alpha_{I1} < \alpha_{I2} < \ldots < \alpha_{IJ}$.

PROOF

The rejection region R of the test consists of all rearrangements $\mathbf{X}^*$ of the sample $\mathbf{X}$ for which $S_I(\mathbf{X}^*)$ greater than $S_I(\mathbf{X})$. Consider first those rearrangements

X*′ consisting of a single pairwise exchange between any two groups. As g is a monotone nondecreasing function, $Pr \{S_I (\mathbf{X}^{*\prime}) \geq S_I (\mathbf{X}) | H_I\} \geq Pr \{S_I (\mathbf{X}^{*\prime}) \geq S_I (\mathbf{X}) | K_I\}$. But by definition, any rearrangement consists of such pairwise exchanges and the theorem follows.

THEOREM 2

Let $\{X_j(\mathbf{s})\}$ be a set of independent observations in a complete balanced experimental design with factors in S such that $X_j(\mathbf{s}) = \Delta \mathbf{s} + \varepsilon_j(\mathbf{s})$, where $\Delta \mathbf{s}$ satisfies Equations B.1 to B.3 and the $\{\varepsilon_j(\mathbf{s})\}$ are a set of independent identically distributed random variables with mean zero. Then the distribution of the statistic $S_{IJ} = \sum_{1 \leq i < i' \leq I} \sum_{1 \leq j < j' \leq J} (X_{..i..j..} + X_{..i'..j'..} - X_{..i'..j..} - X_{..i..j'..})^2$ obtained from synchronized permutations can be used to obtain an exact, unbiased test of a hypothesis concerning a first-order interaction in an N-factor design such as H_{IJ}: $\beta_{ij} = 0$ for $i = 1, \ldots, I; j = 1, \ldots, J$.

PROOF

Let π_{nmlk} denote a synchronized pairwise exchange between the cells of the nth and mth rows and the lth and kth columns of the design, that is, reading from right to left, $\pi_{nmlk} = (1: J; l, k) (1: I; m, n)$ in our previous notation. As can be seen from Figure B.2a, the second exchange affects not only the four cells (l, m), (l, n), (k, m), and (k, n) in the $I \times J$ plane, but the cells in the adjacent rows and columns of the plane. We also see that it suffices to show that the $T_{ij} = X_{ij...} + X_{.i'.j'...} - X_{.i'.j...} - X_{..i.j'..}$ remains invariant with respect to π_{nmlk} when the hypothesis H_{IJ} is true for three of the $2 \times 2 \times \ldots$ subdesigns depicted.

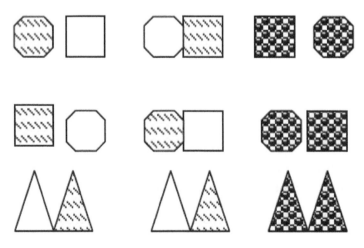

FIGURE B.2a
The same design as in Figure B.1b after a further exchange of column elements. Similar synchronized exchanges have taken place in the hidden dimensions. Note that each combination of shape and pattern corresponds to a distinct first-order interaction term.

From the model in Equation B.1, we see that X_{ij} may be written as the sum of a deterministic portion, which we write as Δ_{ij} and a stochastic portion that we denote by E_{ij}.

Noting that under H_{12}, $\beta_{ij} = 0$ for all i and j, we see that for the $2 \times 2 \times \ldots$ subdesign in the upper-right-hand corner of Figure B.2a, the deterministic portion D_{ij} of $\pi_{nmlk}T_{ij}$ is equal to

$$\Delta_{..i.j..} - \alpha_{1n} + \alpha_{1m} - \alpha_{2l} + \alpha_{2k}$$
$$+ \Delta_{..i'.j'..} - \alpha_{1m} + \alpha_{1n} - \alpha_{2k} + \alpha_{2l}$$
$$- \Delta_{..i'j..} + \alpha_{1m} - \alpha_{1n} + \alpha_{2l} - \alpha_{2k}$$
$$- \Delta_{..i.j'..} + \alpha_{1n} - \alpha_{1m} + \alpha_{2k} - \alpha_{2l}$$

which is equivalent to the deterministic portion of T_{ij}.

The stochastic portion of T_{ij} also remains unchanged by the rearrangement, as it involves a swap of equal numbers of independent identically distributed variables.

For the $2 \times 2 \times \ldots$ subdesign in the lower-right-hand corner of Figure B.2a,

$$\pi_{nmlk}D_{ij} = \Delta_{..i.j..} - \alpha_{1m} + \alpha_{1n} - \alpha_{2k} + \alpha_{2l}$$
$$+ \Delta_{..i'.j'..}$$
$$- \Delta_{..i'j..} - \alpha_{2l} + \alpha_{2k}$$
$$- \Delta_{..i.j'..} - \alpha_{1n} + \alpha_{1m}$$
$$= D_{ij}$$

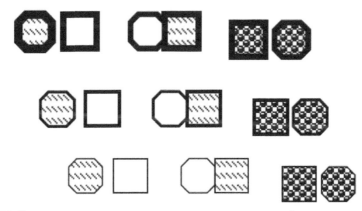

FIGURE B.2b

The same design as in Figure B.2a from a new perspective. This new perspective confirms that similar synchronized exchanges also have taken place in the hidden dimensions. The first row and column of Figure B.2 are displayed at the bottom of this figure. The two additional lines correspond to different levels of a third factor.

For the $2 \times 2 \times \dots$ subdesign formed from the first and second rows and first and third columns of Figure B.2b,

$$\pi_{nmlk} D_{ij} = \Delta_{..i.j..} - \alpha_{1n} + \alpha_{1m} - \alpha_{2l} + \alpha_{2k}$$
$$+ \Delta_{..i'.j'..} - \alpha_{1m} + \alpha_{1n}$$
$$- \Delta_{..i'j..} + \alpha_{1m} - \alpha_{1n} + \alpha_{2l} - \alpha_{2k}$$
$$- \Delta_{..i.j'..} + \alpha_{1n} - \alpha_{1m}$$
$$= D_{ij}$$

All other altered 2×2 subdesigns are similar to one of these three. As synchronized permutations are made up of synchronized pairwise exchanges, the theorem follows.

References

Adams DC, Anthony CD. Using randomization techniques to analyse behavioural data. *Animal Behav.* 1996; 51: 733–738.

Adderley EE. Nonparametric methods of analysis applied to large-scale seeding experiments. *J. Meteorol.* 1961; 18: 692–694

Agresti A. *Categorical data analysis.* 2nd ed. New York: John Wiley & Sons, 2002.

Almasari A, Shukur G. An illustration of the casualty relation between government spending and revenue using wavelet analysis on Finnish data. *J. Appl. Stat.* 2003; 30: 571–584.

Alroy J. Permutation tests for the presence of phylogenetic structure: An editorial. *Syst. Biol.* 1994; 43: 430–437.

Álvarez R, Bécue M, Valencia O. Partial bootstrap in CA: correction of the coordinates. Application to textual data. *JADT* 2006; 8: 43–53.

Aly E-EAA. Simple tests for dispersive ordering. *Stat. Prob. Lett.* 1990; 9: 323–325.

Anderson E. The irises of the Gaspe Peninsula. *Bull. Am. Iris Soc.* 1935; 59: 2–5.

Anderson MJ. Distance-based tests for homogeneity of multivariate dispersions. *Biometrics* 2006; 62: 245–253.

Anderson MJ, Legendre P. An empirical comparison of permutation methods for tests of partial regression coefficients in a linear model. *J. Stat. Comp. Simul.* 1999; 62: 271–303.

Andrews DWK, Buchinsky M. A three-step method for choosing the number of boot-strap repetitions. *Econometrica* 2000; 68: 23–51.

Antretter E, Dunkel D, Haring C. The WHO/EURO multi-centre study of suicidal behaviour: findings of the Austrian research centre in Europe-wide comparison. *Wien. Klin. Wochenschr.* 2000; 112: 955–964.

Arditi D, Pulket T. Predicting the outcome of construction litigation using boosted decision trees. *J. Comp. Civ. Eng.* 2004; 19: 387–393.

Arndt S, Cizadlo T, Andreasen NC, Heckel D, Gold S, Oleary DS. Tests for comparing images based on randomization and permutation methods. *J. Cerebral Blood Flow Metab.* 1996; 16: 1271–1279.

Babu GJ, Feigelson E. *Astrostatistics.* New York: Chapman & Hall, 1996.

Baglivo J, Olivier D, Pagano M. Methods for the analysis of contingency tables with large and small cell counts. *JASA* 1988; 83: 1006–1013.

Baker RD. Two permutation tests of equality of variance. *Stat. Comput.* 1995; 5: 289–296.

Balakrishnan N, Ma CW. A comparative study of various tests for the equality of two population variances. *Stat. Comp. Simul.* 1990; 35: 41–89.

Barker WC, Dayhoff MO. Detecting distant relationships: computer methods and results. In *Atlas of protein sequence and structure*, Dayhoff MO, ed. Silver Spring, MD: National Biomedical Research Foundation, 2006, p. 5.

Barton DE, David FN. Randomization basis for multivariate tests. *Bull. Int. Stat. Inst.* 1961; 39: 455–467.

Bayer RJ, Mabberley DJ, Morton C, Miller CH, Sharma IK, Pfeil BE, Rich S, Hitchcock R, Sykes S. A molecular phylogeny of the orange subfamily (Rutaceae: Aurantioideae) using nine cpDNA sequences. *Am. J. Botany* 2009; 96: 668–685.

Bellec P, Marrelec G, Benali H. A bootstrap test to investigate changes in brain connectivity for functional MRI. *Stat. Sinica* 2008; 18: 1253–1268.

Belmonte M, Yurgelun-Todd D. Permutation testing made practical for functional magnetic resonance image analysis. *IEEE Trans. Med. Imaging* 2001; 20: 243–248.

Benjamini Y, Hochberg Y. Controlling the false discovery rate: a practical and powerful approach to multiple testing. *J. R. Stat. Soc. B* 1995; 57: 289–300.

Benjamini Y, Yekutieli D. The control of the false discovery rate in multiple testing under dependency. *Ann. Stat.* 2001; 29: 1165–1188.

Beran R. Prepivoting to reduce error rate of confidence sets. *Biometrika* 1987; 74: 151–173.

Berger VW. Pros and cons of permutation tests. *Stat. Med.* 2000; 19: 1319–1328.

Berry KJ, Kvamme KL, Mielke PW Jr. Permutation techniques for the spatial analysis of the distribution of artifacts into classes. *Am. Antiquity* 1980; 45: 55–59.

Berry KJ, Kvamme KL, Mielke PW Jr. Improvements in the permutation test for the spatial analysis of the distribution of artifacts into classes. *Am. Antiquity* 1983; 48: 547–553.

Bertail P, Tressou J. Incomplete generalized U-statistics for food-risk assessment. *Biometrics* 2006; 62: 66–74.

Biddle G, Bruton C, Siegel A. Computer-intensive methods in auditing bootstrap difference and ratio estimation. *Auditing J. Practice Theory* 1990; 9: 92–114.

Binder H, Schumacher M. Adapting prediction error estimates for biased complexity selection in high-dimensional bootstrap samples. *Stat. Appl. Genet. Mol. Biol.* 2008; 7: Article 12.

Bishop YMM, Fienberg SE, Holland PW. *Discrete multivariate analysis: theory and practice*. Cambridge, MA: MIT Press, 1975.

Blair C, Higgins JJ, Karinsky W, Krom R, Rey JD. A study of multivariate permutation tests which may replace Hotelling's T test in prescribed circumstances. *Multivariate Behav. Res.* 1994; 29: 141–163.

Blair RC, Troendle JF, Beck RW. Control of familywise errors in multiple endpoint assessments via stepwise permutation tests. *Stat. Med.* 1996; 15: 1107–1121.

Bollen KA, Stine RA. Bootstrapping goodness-of-fit measures in structural equation models. In *Testing structural equation models*, Bollen KA, Long JS, eds. Beverly Hills, CA: Sage Publications, 1993, pp. 111–115.

Bond J, Michailidi SG. Interactive correspondence analysis in a dynamic object-oriented environment. *J. Stat. Soft.* 1997; 2: 1–30.

Boschloo RD. Raised conditional level of significance for the 2×2 table when testing the equality of two probabilities. *Stat. Neer.* 1970; 24: 1–35.

Bradley JV. *Distribution free statistical tests*. Englewood Cliffs, NJ: Prentice Hall, 1968.

Braun TM, Feng Z. Optimal permutation tests for the analysis of group randomized trials. *J. Am. Stat. Assoc.* 2001; 96: 1424–1432.

Breiman L, Friedman JH, Olshen RA, Stone CJ. *Classification and regression trees*. Monterey, CA: Wadsworth and Brooks, 1984.

Brey T. Confidence limits for secondary prediction estimates. *Mar. Biol.* 1990; 106: 503–508.

Bross IDJ. Taking a covariable into account. *JASA* 1964; 59: 725–736.

Brown MB, Forsythe AB. Robust tests for equality of variances. *JASA* 1974; 69: 364–367.

Bryan J. Problems in gene clustering based on gene expression data. *J. Mult. Anal.* 2004; 90: 44–66.

Bryant EH. Morphometric adaptation of the housefly, *Musa domestica* L, in the United States. *Evolution* 1977; 31: 580–956.

Buckinx W, Van den Poel D. Customer base analysis: partial defection of behaviourally loyal clients in a non-contractual FMCG retail setting. *Eur. J. Oper. Res.* 2005; 164: 252–268.

Bühlmann P, Hothorn T. Boosting algorithms: regularization, prediction and model fitting. *Stat. Sci.* 2007; 22: 477–505.

Cadarso-Suárez C, Roca-Paridiñas J, Molrnberghs G, Faes C, Nácher V, Ojeda S, Acuña C. Flexible modelling of neuron firing rates across different experimental conditions. *JRSS C* 2006; 55: 431–447.

Cade B, Hoffman H. Differential migration of blue grouse in Colorado. *Auk* 1993; 110: 70–77.

Cade BS, Richards JD. Permutation tests for least absolute deviation regression. *Biometrics* 1996; 52: 886–902.

Cade BS, Richards JD. A permutation test for quantile regression. *J. Agric. Biol. Environ. Stat.* 2006; 11: 106–126.

Canty AJ, Davison AC, Hinkley DV, Ventura V. Bootstrap diagnostics. http://www.stat.cmu.edu/www/cmu-stats/tr/tr726/tr726.html. 2000.

Cardot H, Prchal L, Sarda P. No effect and lack-of-fit permutation tests for functional regression. *Comput. Stat.* 2007; 2: 371–390.

Carlson JM, Heckerman D, Shani G. Estimating false discovery rates for contingency tables. No. MicroSoftResearch-TR-2009-53. May 2009.

Chapelle JP, Albert A, Smeets JP, Heusghem C, Kulberts HE. Effect of the hyptoglobin phenotype on the size of a myocardial infarction. *NEJM.* 1982; 307:457-463.

Chen Y, Liu JS. Sequential Monte Carlo methods for permutation tests on truncated data. *Stat. Sinica* 2007; 17: 857–872.

Chernick MR. *Bootstrap methods: a practitioner's guide.* 2nd ed. New York: Wiley, 2008.

Chernick MR, Liu CY. The saw-toothed behavior of power versus sample size and software solutions: single binomial proportion using exact methods. *Am. Stat.* 2002; 56: 149–155.

Choi KC, Nam KH, Park DH. Estimation of capability index based on bootstrap method. *Microelectron. Reliab.* 1996; 36: 256–259.

Clark LA, Pregibon D. Tree-based models. In *Statistical models in S*, Chambers JM, Hastie TJ, eds. New York: Chapman & Hall, 1992.

Coakley KJ. Bootstrap method for nonlinear filtering of EM-ML reconstructions of PET images. *Int. J. Imaging Syst. Technol.* 1996; 7: 54–61.

Cohen A, Sackrowitz HB. Methods of reducing loss of efficiency due to discreteness of distributions. *Stat. Meth. Med. Res.* 2003; 12: 23–36.

Comley JW, Dowe DL. General Bayesian networks and asymmetric languages. Paper presented at Proceedings of the 2nd Hawaii International Conference on Statistics and Related Fields, June 5–8, 2003.

Conover WJ, Johnson ME, Johnson MM. Comparative study of tests for homogeneity of variances: with applications to the outer continental shelf bidding data. *Technometrics* 1981; 23: 351–361.

Cressie N, Read T. Multinomial goodness of-fit tests. *JRSS B* 1984; 46: 440–464.

Davis AW. On the effects of moderate non-normality on Roy's largest root test. *JASA* 1982; 77: 896–900.

Davis RE, Elder K. Application of classification and regression trees: selection of avalanche activity indices at Mammoth Mountain. In *Proceedings of the International Snow Science Workshop in Snowbird Utah*, 1994, pp. 285–294.

Davison AC, Hinkley DV. *Bootstrap Methods and Their Application.* Cambridge Univeristy Press. 1997.

Daw NC, Arnold JT, Abushullaih BA, Stenberg PE, White MM, Jayawardene D, Srivastava DK, Jackson CWA. Single intravenous dose murine megakaryocyte growth and development factor potently stimulates platelet production challenging the necessity for daily administration. *Blood* 1998; 91: 466–474.

de la Calleja J, Fuentes O. Automated classification of galaxy images. In *Proceedings of the Eight International Conference on Knowledge-Based Intelligent Information and Engineering Systems (KES)*, Wellington, New Zealand, September 2004. Lecture Notes in Artificial Intelligence 3215.

Dee CR, Rankin JA, Burns CA. Using scientific evidence to improve hospital library services. *B. Med. Libr. Assoc.* 1998; 86: 301–306.

Derado G, Mardia K, Patrangenaru V, Thompson H. A shape-based glaucoma index for tomographic images. *J. Appl. Stat.* 2004; 31: 1241–1248.

Dette H, Podoloskij M, Vetter M. Estimation of integrated volatility in continuous time financial models with applications to goodness-of-fit testing. *Scand. J. Stat.* 2006; 33: 259–278.

Diaconis P, Graham R, Holmes S. Statistical problems involving permutations with restricted positions. In *State of the art in probability and statistics* (Festschrift in honor of William van Zwet). 36 of IMS Lecture Notes. Beachwood, OH: IMS, 2001, pp. 195–222.

DiCiccio TJ, Romano J. A review of bootstrap confidence intervals (with discussions). *JRSS B* 1988; 50: 163–170.

Do KA, Hall P. On importance resampling for the bootstrap. *Biometrika* 1991; 78: 161–167.

Dobbin K, Simon R. Sample size determination in microarray experiments for class comparison and prognostic classification. *Biostatistics* 2005; 6: 27–38.

Dolnicar S, Leisch F. Evaluation of structure and reproducibility of cluster solutions using the bootstrap. *Marketing Lett.* 2010; 21: 83–101.

Dudoit S, Fridlyand J. Classification in microarray experiments. In *Statistical analysis of gene expression microarray Data.* London: Chapman & Hall, 2003, pp. 93–158.

Dupont WD. Sensitivity of Fisher's exact test to minor perturbations in 2×2 contingency tables. *Stat. Med.* 1986; 5: 629–635.

Eden T, Yates F. On the validity of Fisher's z test when applied to an actual sample of nonnormal data. *J. Agric. Sci.* 1933; 23: 6–16.

Edgington ES, Bland BH. Randomization tests: application to single-cell and other single-unit neuroscience experiments. *J. Neurosci. Meth.* 1993; 47: 169–177.

Edgington ES, Onghena P. *Randomization tests.* 4th ed. New York: CRC, 2007.

Efron B. Bootstrap methods: another look at the jackknife. *Ann. Stat.* 1979; 7: 1–26.

Efron B. *The jackknife, the bootstrap and other resampling plans.* Philadelphia: SIAM, 1982.

Efron B. Better bootstrap confidence intervals (with disc.) *JASA* 1987; 82, 171–200.

Efron B. Bootstrap confidence intervals: good or bad? (with discussion). *Psychol. Bull.* 1988; 104: 293–296.

Efron B. Six questions raised by the bootstrap. In *Exploring the limits of the bootstrap*, LePage R, Billard L, eds. New York: Wiley, 1992.

Efron B, Tibshirani R. Bootstrap measures for standard errors, confidence intervals, and other measures of statistical accuracy. *Stat Sci.* 1986; 1: 54–77.

Efron B, Tibshirani R. *An introduction to the bootstrap.* New York: Chapman & Hall, 1993.

Elderton EM, Barrington A, Jones HG, Lamotte EM, Laski HJ, Pearson K. *On the correlation of fertility with social value. A cooperative study.* Eugenics Laboratory Memoirs XVIII. University of London, 1913.

Faris PD, Sainsbury RS. The role of the pontis oralis in the generation of RSA activity in the hippocampus of the guinea pig. *Psychol. Behav.* 1990; 47: 1193–1199.

Farrar DA, Crump KS. Exact statistical tests for any cocarcinogenic effect in animal assays. *Fund. Appl. Toxicol.* 1988; 11: 652–663.

Farrar DA, Crump KS. Exact statistical tests for any cocarcinogenic effect in animal assays. II. Age adjusted tests. *Fund. Appl. Toxicol.* 1991; 15: 710–721.

Feinstein AR. Clinical biostatistics XXIII. The role of randomization in sampling, testing, allocation, and credulous idolatry (part 2). *Clin. Pharm.* 1973; 14: 989–1019.

Fernández de Castro B, Guillas S, González-Manteiga W. Functional samples and bootstrap for predicting sulphur dioxide levels. *Technometrics* 2005; 47: 212–222.

Fischler MA, Bolles RC. Random sample consensus: a paradigm for model fitting with applications to image analysis and automated cartography. *Commun. ACM* 1981; 24: 381–395.

Fisher RA. *Design of experiments.* New York: Hafner, 1935.

Fisher RA. The use of multiple measurements in taxonomic problems. *Ann. Eugenics* 1936; 7: 179–188.

Ford RD, Colom LV, Bland BH. The classification of medial septum-diagonal band cells as theta-on or theta-off in relation to hippocampal EEG states. *Brain Res.* 1989; 493: 269–282.

Foutz RN, Jensen DR, Anderson GW. Multiple comparisons in the randomization analysis of designed experiments with growth curve responses. *Biometrics* 1985; 41: 29–37.

Frank D, Trzos RJ, Good P. Evaluating drug-induced chromosome alterations. *Mutation Res.* 1978; 56: 311–317.

Freeman GH, Halton JH. Note on an exact treatment of contingency, goodness of fit, and other problems of significance. *Biometrika* 1951; 38: 141–149.

Gabriel KR. Some statistical issues in weather experimentation. *Commun. Stat. A* 1979; 8: 975–1015.

Gail M, Mantel N. Counting the number of rxc contingency tables with fixed marginals. *JASA* 1977; 72: 859–862.

Gart J. Pont and interval estimation of the common odds ratio in the combination of 2 × 2 tables with fixed margins. *Biometrika* 1970; 57: 471–475.

Gart JJ. Statistical methods in cancer research: the design and analysis of long term animal experiments. Vol. III. Lyon: IARC Scientific Publications, 1986.

Gastwirht JL. Statistical reasoning in the legal setting. *Am. Stat.* 1992; 46: 55–69.

Gavrilov Y, Benjamini Y, Sarkar SK. An adaptive step-down procedure with proven FDR control under independence. *Ann. Stat.* 2009; 37: 619–629.

Geisser S. The predictive sample reuse method with applications. *JASA* 1975; 70: 320–328.

Giancristofaro RA, Bonnini S. Some new results on univariate and multivariate permutation tests for ordinal categorical variables under restricted alternatives. *Stat. Meth. Appl.* 2009; 18: 221–236.

Gine E, Zinn J. Necessary conditions for a bootstrap of the mean. *Ann. Stat.* 1989; 17: 684–691.

Glass AG, Mantel N, Gunz FW, Spears GFS. Time-space clustering of childhood leukemia in New Zealand. *J. Natl. Cancer Inst.* 1971; 47: 329–336.

Gleason JR. Algorithms for balanced bootstrap simulations. *Am. Stat.* 1988; 42: 263–266.

Gliddentracey CE, Parraga MI. Assessing the structure of vocational interests among Bolivian university students. *J. Vocational Behav.* 1996; 48: 96–106.

Goeman JJ, Bühlmann P. Analyzing gene expression data in terms of gene sets: methodological issues. *Bioinformatics* 2007; 23: 980–987.

Goldberg P, Leffert F, Gonzales M, Gorgenola I, Zerbe GO. Intraveneous aminophylline in asthma: a comparison of two methods of administration in children. *Am. J. Dis. Children* 1980; 134: 12–18.

Gong G. Cross-validation, the jackknife and the bootstrap: excess error in forward logistic regression. *JASA* 1986: 81: 108–113.

Gonzalez JR, Garcia-Moro C, Dahinten S, Hernandez M. Origin of Fueguian-Patagonians: an approach to population history and structure using R matrix and matrix permutation methods. *Am. J. Hum. Biol.* 2002; 14: 308–320.

Good PI. Detection of a treatment effect when not all experimental subjects respond to treatment. *Biometrics* 1979; 35: 483–489.

Good PI. Most powerful tests for use in matched pair experiments when data may be censored. *J. Stat. Comp. Simul.* 1991; 38: 57–63.

Good PI. *Permutation tests.* 2nd ed. New York: Springer Verlag, 2000.

Good PI. *Applying statistics in the courtroom.* London: Chapman & Hall, 2001.

Good PI. Extensions of the concept of exchangeability and their applications. *J. Modern Appl. Stat. Meth.* 2002; 1: 243–247.9

Good PI. *Permutation tests.* 3rd ed. New York: Springer Verlag, 2005a.

Good PI. *Introduction to statistics via resampling methods and R.* New York: Wiley, 2005b.

Good PI. Robustness of Pearson correlation. http://interstat.statjournals.net/ YEAR/2009/articles/0906005.pdf. 2009.

Good PI, Hardin J. *Common errors in statistics.* 34th ed. New York: Wiley, 2009.

Good PI, Lunneborg L. Limitations of the analysis of variance. The one-way design. *J. Modern Appl. Stat. Meth.* 2006; 5: 41–43.

Good PI, Xie F. Analysis of a crossover clinical trial by permutation methods. *Contemp. Clin. Trials* 2008; 29: 565–568.

Goodman L, Kruskal W. Measures of association for cross-classification. *JASA* 1954; 49: 732–764.

Graubard BI, Korn EL. Choice of column scores for testing independence in ordered 2 by K contingency tables. *Biometrics* 1987; 43: 471–476.

Grossman DC, Cummings P, Koepsell TD, et al. Firearm safety counseling in primary care pediatrics: a randomized controlled trial. *Pediatrics* 2000; 106: 22–26.

Guilbaud O. Exact comparisons of means and within-subject variances in 2 × 2 crossover trials. *Drug Inform. J.* 1999; 33: 455–469.

Hall P. On the bootstrap and confidence intervals. *Ann. Stat.* 1986; 14: 1431–1452.

Hall P. Theoretical comparison of bootstrap confidence intervals (with discussion). *Ann. Stat.* 1988; 16: 927–985.

Hall P, Hart JD. Bootstrap test for difference between means in nonparametric regression. *JASA* 1990; 85: 1039–1049.

Hall P, Martin MA. On bootstrap resampling and iteration. *Biometrika* 1988; 75: 661–671.

Halter JH. A rigorous derivation of the exact contingency formula. *Proc. Cambridge Phil. Soc.* 1969; 65: 527–530.

Hammoudi DS, Lee SSF, Madison A, Mirabella G, Buncic JR, Logan WJ, Snead OC, Westall CA. Reduced visual function associated with infantile spasms in children on Vigabatrin therapy. *Invest. Ophthalmol. Visual Sci.* 2005; 46: 514–520.

Hartigan JA. Using subsample values as typical values. *JASA* 1969; 64: 1303–1317.

Hartigan JA. Error analysis by replaced samples. *J. R. Stat. Soc. B* 1971; 33: 98–110.

Hayasaka S, Nichols TE. Combining voxel intensity and cluster extent with permutation test framework. *Neuroimage* 2004; 23: 54–63.

Hayden D, Lazar P, Schoenfeld D. Assessing statistical significance in microarray experiments using the distance between microarrays. *PLoS ONE* 2009; 4(6): e5838. doi:10.1371/journal.pone.0005838.

Hettmansperger TP. *Statistical inference based on ranks.* New York: Wiley, 1984.

Higgins JJ, Noble W. A permutation test for a repeated measures design. *Appl. Stat. Agric.* 1993; 5: 240–525.

Hinkley DV, Shi S. Importance sampling and the nested bootstrap. *Biometrika* 1989; 76: 435–446.

Hisdal H, Stahl K, Tallaksen LM, et al. Have streamflow droughts in Europe become more severe or frequent? *Int. J. Climatol.* 2001; 21: 317–321.

Hjorth JSU. *Computer intensive statistical methods: validation, model selection and bootstrap.* New York: Chapman & Hall, 1994.

Hoel DG, Walburg HE. Statistical analysis of survival experiments. *J. Natl. Cancer Inst.* 1972; 49: 361–72.

Hoff A. Bootstrapping Malmquist indicies for Danish seiners in the North Sea and Skagerrak. *J. Appl. Stat.* 2006; 33: 891–907.

Hollander M, Sethuraman J. Testing for agreement between two groups of judges. *Biometrika* 1978; 65: 403–412.

Homer C, Dewitz J, Fry J, Coan M, Hossain N, Larson C, Herold N, McKerrow AJ, VanDriel JN, Wickham J. Completion of the 2001 National Land Cover Database for the Conterminous United States. *Photogram. Eng. Remote Sensing* 2007; 73: 337–341.

Howard M (pseudonym for Good P). Randomization in the analysis of experiments and clinical trials. *Am. Lab.* 1981; 13: 98–102.

Huang A, Rungao J, Robinson J. Robust permutation tests for two samples. *J. Stat. Plan. Infer.* 2009; 139: 2631–2642.

Hubert L, Arabie P. Comparing partitions. *J. Classification* 1985; 2: 193–218.

Hubert LJ. Combinatorial data analysis: association and partial association. *Psychometrika* 1985; 50: 449–467.

Hubert LJ, Schultz J. Quadratic assignment as a general data analysis strategy. *Br. J. Math. Stat. Psychol.* 1976; 29: 190–241.

Ilgen MA, Downing K, Zivin K, Hoggatt KJ, Kim HM, Ganoczy D, Austin KL, McCarthy JF, Patel JM, Valenstein M. Exploratory data mining analysis identifying subgroups of patients with depression who are at high risk for suicide. *J. Clin. Psychiatry* 2009; 70: 1495–1500.

Ingenbleek JF. Tests simultanes de permutation des rangs pour bruit-blanc multivarie. *Stat. Anal. Donnees.* 1981; 6: 60–65.

Jackson DA. Ratios in aquatic sciences: statistical shortcomings with mean depth and the morphoedaphic index. *Can. J. Fisheries Aquat. Sci.* 1990; 47: 1788–1795.

Jagers P. Invariance in the linear model: An argument for chi-square and F in nonnormal situations. *Mathematische Operations Forschung und Statistik,* 80; 11: 455–464.

Janssen I, Stebbings JH, Essling MA, Dimino KM, Rodgers DL. Prediction of RN-222 from topography in Pennsylvania, *Health Physics* 1991; 61. 775–783.

Jiang W, Simon R. A comparison of bootstrap methods and an adjusted bootstrap approach for estimating the prediction error in microarray classification. *Stat. Med.* 2007; 26: 5320–5334.

Johns MV Jr. Importance sampling for bootstrap confidence intervals. *JASA* 1988; 83: 709–714.

Johnson WD, Mercante DE. Analyzing multivariate data in crossover designs using permutation tests. *J. Biopharm. Stat.* 1996; 63: 327–342.

Jones G, Wortberg M, Kreissig SB, Hammock BD, Rocke DM. Application of the bootstrap to calibration experiments. *Anal. Chem.* 1996; 68: 763–770.

Jones HL. Investigating the properties of a sample mean by employing random subsample means. *JASA* 1956; 51: 54–83.

Jorde LB, Rogers AR, Barnshad M, Watkins WS, Krakowiak P, Sung S, Kere J, Harpending HC. Microsatellite diversity and the demographic history of modern humans. *Proc. Nat. Acad. Sci., USA.* 1997; 94: 3100–3103.

Kalbfleisch JD, Prentice RL. *The statistical analysis of failure time data.* New York: John Wiley & Sons, 1980.

Karaolis M, Moutiris J, Hadjipanayi D, Pattchis C. Assessment of the risk factors of coronary heart events based on data mining with decision trees. *IEEE Trans. Inf. Technol. Biomed.* 2010; 14: 559–566.

Karlin S, Williams PT. Permutation methods for the structured exploratory data analysis (SEDA) of familial trait values. *Am. J. Hum. Genet.* 1984; 36: 873–898.

Kass GV. An exploratory technique for investigating large quantities of categorical data. *J. Appl. Stat.* 1980; 29: 119–127.

Kempthorne O. The randomization theory of experimental inference. *JASA* 1955; 50: 946–967.

Kempthorne O. Why randomize? *J. Stat. Prob. Infer.* 1977; 1: 1–26.

Kennedy PE. Randomization tests in econometrics. *J. Business and Economic Statist.* 1995; 13: 85–95.

Kerr MK, Churchill GA. Bootstrapping cluster analysis: assessing the reliability of conclusions from microarray experiments. *Proc. Natl. Acad. Sci. USA* 2001; 98: 8961–8965.

Kim H, Loh W-Y. Classification trees with unbiased multiway splits. *JASA* 2001; 96: 589–604.

Kingsford C, Salzberg SL. What are decision trees? *Nature Biotechnol.* 2008; 26: 1011–1013.

Klauber MR. Two-sample randomization tests for space-time clustering. *Biometrics* 1971; 27: 129–142.

Knight K. On the bootstrap of the sample mean in the infinite variance case. *Ann. Stat.* 1989; 17: 1168–1173.

Kobler A, Adamič M. Brown bears in Slovenia: identifying locations for construction of wildlife bridges across highways. In *Proceedings of the Third International Conference on Wildlife Ecology and Transportation*, Evink GL, Garrett P, Zeigler D, eds., Missoula, MT, September 13–16, 1999. Tallahassee: Florida Department of Transportation, 1999, pp. 29–38.

Kong SW, Pu WT, Park PJ. A multivariate approach for integrating genome-wide expression data and biological knowledge. *Bioinformatics* 2006; 22: 2373–2380.

Koziol JA, Maxwell DA, Fukushima M, Colmer A, Pilch YHA. Distribution-free test for tumor-growth curve analyses with applications to an animal tumor immunotherapy experiment. *Biometrics* 1981; 37: 383–390.

Krewski D, Brennan J, Bickis M. The power of the Fisher permutation test in 2 by k tables. *Commun. Stat. B* 1984; 13: 433–448.

Kuo W-J, Chang R-F, Moon WK, Lee CC, Chen D-R. Computer-aided diagnosis of breast tumors with different US systems. *Acad. Radiol.* 2002; 9: 793–799.

Kwon H-H, Moon Y-I. Improvement of overtopping risk evaluations using probabilistic concepts for existing dams. *Stochastic Environ. Res. Risk Assess.* 2006; 20: 223–237.

Lachin JM. Properties of sample randomization in clinical trials. *Contr. Clin. Trials* 1988a; 9: 312–326.

Lachin JM. Statistical properties of randomization in clinical trials. *Contr. Clin. Trials* 1988b; 9: 289–311.

Ladanyi A, Sher AC, Herlitz A, Bergsrud DE, Kraeft SK, Kepros J, McDaid G, Ferguson D, Landry ML, Chen LB. Automated detection of immunofluorescently labeled cytomegalovirus-infected cells in isolated peripheral blood leukocytes using decision tree analysis. *Cytometry* 2004; 58A: 147–156.

Lahiri SN. *Resampling methods for dependent data.* New York: Springer, 2003.

Laitenberger O, Atkinson C, Schlich M, et al. An experimental comparison of reading techniques for defect detection in UML design documents. *J. Syst. Software* 2000; 53: 183–204.

Landau S, Ellison-Wright IC, Bullmore ET. Tests for a difference in timing of physiological response between two brain regions measured by using functional magnetic resonance imaging. *JRSS C* 2004; 53: 63–82.

Langeheine R, Pannekoek J, van de Pol F. Bootstrapping goodness-of-fit measures in categorical data analysis. *Sociol. Meth. Res.* 1996; 24: 492–516.

Lanyon SM. Jackknifing and bootstrapping: important "new" statistical techniques for ornithologists. *Auk* 1987; 104: 144–146.

Lee D. Analysis of phase-locked oscillations in multi-channel single-unit spike activity with wavelet cross-spectrum. *J. Neurosci. Meth.* 2002; 115: 67–75.

Lee SMS, Pun MC. On m out of n bootstrapping for nonstandard m-estimation with nuisance parameters. *JASA* 2006; 101: 1185–1197.

Lehmann EL. *Testing statistical hypotheses.* 2nd ed. New York: John Wiley & Sons, 1986.

Lim TS, Loh WY, Shih Y-S. A comparison of prediction accuracy complexity and training time of thirty-three old and new classification algorithms. *Machine Learning J.* 2000; 40: 203–228.

Lin S, Rogers JA, Hsu JC. A confidence-set approach for finding tightly linked genomic regions. *Am. J. Hum. Genet.* 2001; 68: 1219–1228.

Linting M, Meulman JJ, Groenen PJF, Van der Kooij JJ. Stability of nonlinear principal components analysis: an empirical study using the balanced bootstrap. *Psychol. Meth.* 2007; 12: 359–379.

Liu RY, Singh K. Moving blocks jackknife and bootstrap capture weak dependence. In *Exploring the limits of bootstrap*, LePage R, Billard L, eds. New York: John Wiley, 1992, pp. 225–248.

Loh W-Y. Calibrating confidence coefficients. *JASA* 1987; 82: 155–162.

Loh W-Y. Bootstrap calibration for confidence interval construction and selection. *Stat. Sinica* 1991; 1: 479–495.

Loh WY, Shih Y-S. Split selection methods for classification trees. *Stat. Sinica* 1997; 7: 815–840.

Loh W-Y, Vanichsetakul N. Tree-structured classification via generalized discriminant analysis (with discussion). *JASA* 1988; 83: 715–728.

Loughin TM, Noble W. A permutation test for effects in an unreplicated factorial design. *Technometrics* 1997; 39: 180–190.

Luger R. Exact permutation tests for non-nested non-linear regression models. *J. Economics* 2006 ; 133: 513–529.

Lynden-Bell D. A method of allowing for a known observational selection in small samples applied to 3CR quasars. *Monogr. Natl. R. Astrophys. Soc.* 1971; 155: 95–118.

Mackert JR Jr., Twiggs SW, Russell CM, Williams AL. Evidence of a critical leucite particle size for microcracking in dental porcelains. *J. Dent. Res.* 2001; 80: 1574–1579.

Majeed AW, Troy G, Nicholl JP, Smythe A, Reed MWR, Stoddard CJ, Peacock J, Johnson AG. Randomized prospective single-blind comparison laparoscopic versus small-incision cholecystectomy. *Lancet* 1996; 347: 989–994.

Makinodan T, Albright JW, Peter CP, Good PI, Hedrick ML. Reduced humoral activity in long-lived mice. *Immunology* 1976; 31: 400–408.

Manly BFJ. The comparison and scaling of assessment in several subjects. *Applied Statistics.* 1988; 37: 385–395.

Manly BFJ. *Randomization, bootstrap and Monte Carlo methods in biology.* 3rd ed. London: Chapman & Hall, 2006.

Mantel N. The detection of disease clustering and a generalized regression approach. *Cancer Res.* 1967; 27: 209–220.

Mapleson WW. The use of GLIM and the bootstrap in assessing a clinical trial of two drugs. *Stat. Med.* 1986; 5: 363–374.

Marascuilo LA, McSweeny M. *Nonparametric and distribution-free methods for the social sciences.* Monterey CA: Brooks/Cole, 1977.

Marcus LF. Measurement of selection using distance statistics in prehistoric orangutan pongo pygamous palaeosumativens. *Evolution* 1969; 23: 301.

Maritz JS. *Distribution free statistical methods.* 2nd ed. London: Chapman & Hall, 1996.

Marron JS. A comparison of cross-validation techniques in density estimation. *Ann. Stat.* 1987; 15: 152–162.

Martin MA. On bootstrap iteration for converge correction in confidence intervals. *JASA* 1990; 85: 1105–1108.

Maxwell SE, Cole DA. A comparison of methods for increasing power in randomized between-subjects designs. *Psychol. Bull.* 1991; 110: 328–337.

McCarthy PJ. Pseudo-replication: half samples. *Rev. Int. Stat. Inst.* 1969; 37: 239–264.

McDonald LL, Davis BM, Miliken GA. A nonrandomized unconditional test for comparing two proportions in 2 × 2 contingency tables. *Technometrics* 1977; 19: 145–158.

McIntyre LM, Martin ER, Simonsen KL, Kaplan NL. Circumventing multiple testing: a multilocus Monte Carlo approach to testing for association. *Genet. Epidemiol.* 2000; 19: 18–29.

McKinney PW, Young MJ, Hartz A, Bi-Fong Lee M. The inexact use of Fisher's exact test in six major medical journals. *JAMA* 1989; 261: 3430–3433.

McQueen G. Long-horizon mean-reverting stock prices revisited. *J. Financial Quant. Anal.* 1992; 27: 1–17.

Mehta CR, Patel NR. A network algorithm for performing Fisher's exact test in rxc contingency tables. *JASA* 1983; 78: 427–434.

Mehta CR, Patel NR, Gray R. On computing an exact confidence interval for the common odds ratio in several 2 × 2 contingency tables. *JASA* 1985; 80: 969–973.

Mehta CR, Patel NR, Senchaudhuri P. Importance sampling for estimating exact probabilities in permutational inference. *JASA* 1988; 83: 999–1005.

Mielke PW. Non-metric statistical analysis: some metric alternatives. *J. Stat. Plan. Infer.* 1986; 13: 377–387.

Mielke PW Jr, Berry KJ. Multivariate tests for correlated data in completely randomized designs. *J. Educ. Behav. Stat.* 1999; 24: 109–131.

Miller RA, Bookstein F, Vandermeulen J, Engle S, Kim J, Mullins L, Faulkner J. Candidate biomarkers aging—age sensitive indexes of immune and muscle function covary in genetically heterogenous mice. *J. Gerontology A. Biol. Sci. Med. Sci.* 1997; 52: B39–B47.

Miller RG. Jackknifing variances. *Ann. Math. Stat.* 1968; 39: 567–582.

Milligan GW, Cooper MC. An examination of procedures for determining the number of clusters in a data set. *Psychometrika* 1985; 50: 159–179.

Mitchell-Olds T. Analysis of local variation in plant size. *Ecology* 1987; 68: 82–87.

Molinaro AM, Simon R, Pfeiffer RM. Prediction error estimation: a comparison of resampling methods. *Bioinformatics* 2005; 21: 3301–3307.

Mooney CZ. Bootstrap statistical inference: examples and evaluation for political science. *Am. J. Pol. Sci.* 1996; 40: 570–602.

Morgan JN, Sonquist JA. Problems in the analysis of survey data and a proposal. *JASA* 1963; 58: 415–434.

Mosteller F, Tukey JW. *Data analysis and regression: a second course in statistics.* Menlo Park, CA: Addison-Wesley, 1977.

Nelson LS. A randomization test for ordered alternatives. *J. Qual. Technol.* 1992; 24: 51–53.

Nguyen TT. A generalization of Fisher's exact test in pxq contingency tables using more concordant relations. *Commun. Stat. B* 1985; 14: 633–645.

Nichols TE, Holmes AP. Nonparametric permutation tests for functional neuroimaging: a primer with examples. *Hum. Brain Mapping* 2001; 15: 1–25.

Nistér D, Naroditsky O, Bergen J. Visual odometry for ground vehicle applications. *J. Field Robotics* 2006; 23: 3–20.

Noreen E. *Computer intensive methods for testing hypotheses.* New York: John Wiley & Sons, 1989.

North BV, Curtis D, Cassell PG, Hitman GA, Sham PC. Assessing optimal neural network architecture for identifying disease-associated multi-marker genotypes using a permutation test, and application to calpain 10 polymorphisms associated with diabetes. *Ann. Hum. Genet.* 2003; 67: 348–356.

Nurminen M. Prognostic models for predicting delayed onset of renal allograft function. *Internet J. Epidemiol.* 2003; 1: 1.

Oehler F, Rutherford JC, Coco G. The use of machine learning algorithms to design a generalized simplified denitrification model. *Biogeosci. Discuss.* 2010; 7: 3311–3332.

O'Gorman TW. An adaptive test of a subset of regression coefficients using permutations of residuals, *J. Stat. Comp. Simul.* 2006; 76: 1095–1105.

Ojala M, Garriga GC. Permutation tests for studying classifier performance. *J. Mach. Learning Res.* 2010; 11: 1833–1863.

Orlowski LA, Grundy WD, Mielke PW, Schumm S. Geological applications of multi-response permutation procedures. *Math. Geol.* 1993; 25: 483–500.

O'Sullivan F, Whitney P, Hinsnelwood MM, Hauser ER. Analysis of repeated measurement experiments in endocrinology. *J. Animal Science.* 1989; 59: 1070–1079.

Pampoulie C, Morand S. Nonrandom association patterns in parasite infections caused by the host life cycle: empirical evidence from *Kudoa caguensis* (Myxosporea) and *Aphalloides coelomicola* (Trematoda). *J. Parasitol.* 2002; 88: 817–819.

Park ES, Spiegelman CH, Henry RC. Bilinear estimation of pollution source profiles and amounts by using multivariate receptor models. *Environ. Metrics* 2002; 13: 775–798.

Passing H. Exact simultaneous comparisons with controls in an rxc contingency table. *Biometrical J.* 1984; 26: 643–654.

Patefield M. Conditional and exact tests in crossover trials. *J. Biopharm. Stat.* 2000; 101: 109–129.

Patefield WM. Exact tests for trends in ordered contingency tables. *Appl. Stat.* 1982; 31: 32–43.

Patil CHK. Cochran's Q test: exact distribution. *JASA* 1975; 70: 186–189.

Pearson ES. Some aspects of the problem of randomization. *Biometrika* 1937; 29: 53–64.

Penninckx W, Hartmann C, Massart DL, Smeyersverbeke J. Validation of the calibration procedure in atomic absorption spectrometric methods. *J. Anal. Atomic Spectrom.* 1996; 11: 237–246.

Peritz E. Exact tests for matched pairs: studies with covariates. *Commun. Stat. A* 1982; 11: 2157–2167 (errata 12: 1209–1210).

Peritz E. Modified Mantel-Haenszel procedures for matched pairs. *Commun. Stat. A* 1985; 14: 2263–2285.

Perlich C, Provost F, Simonoff JS. Tree induction vs. logistic regression: a learning-curve analysis. *J. Machine Learning Res.* 2003; 4: 211–255.

Pesarin F. On a nonparametric combination method for dependent permutation tests with applications. *Psychother. Psychosom.* 1990; 54: 172–179.

Pesarin F. A nonparametric combination method for dependent permutation tests with application to some problems with repeated measures. In *Industrial statistics*, Kitsos CP, Edler L, eds. Heidelberg: Physics-Verlag, 1997, pp. 259–68.

Pesarin F. *Multivariate permutation tests.* New York: Wiley, 2001.

Pesarin F, Salmaso L. *Permutation tests for complex data: theory, applications and software.* New York: Wiley, 2010.

Petrondas DA, Gabriel RK. Multiple comparisons by rerandomization tests. *JASA* 1983; 78: 949–957.

Phipps MC. Small samples and the tilted bootstrap. *Theory Stochastic Processes* 1997; 19: 355–362.

Pitman EJG. Significance tests which may be applied to samples from any population. *R. Stat. Soc. Suppl.* 1937; 4: 119–130, 225–232.

Pitman EJG. Significance tests which may be applied to samples from any population. Part III. The analysis of variance test. *Biometrika* 1938; 29: 322–335.

Plackett RL, Hewlett PS. A unified theory of quantal responses to mixtures of drugs. The fitting to data of certain models for two non-interactive drugs with complete positive correlation of tolerances. *Biometrics* 1963; 19: 517–531.

Pollard E, Lackland KH, Rothrey P. The detection of density dependence from a series of annual censuses. *Ecology* 1987; 68: 2046–2055.

Ponton D, Copp GH. Early dry-season community structure and habitat use of young fish in tributaries of River Sinnamary (French Guiana, South America) before and after hydrodam operation. *Environ. Biol. Fishes.* 1997; 50: 235–256.

Potter D, Griffiths D. Omnibus permutation tests of the overall null hypothesis in datasets with many covariates. *J. Biopharm. Stat.* 2006; 16: 327–341.

Puri ML, Sen PK. On a class of multivariate, multisample rank-order tests. *Sankyha Ser. A* 1966; 28: 353–376.

Qin J, Liang K-Y. Hypothesis testing in a mixture case-control model. *Biometrics* 2011; 67: 182–193.

Quinlan JR. Bagging, boosting, and C4.5. In *Proceedings of the 13th National Conference on Artificial Intelligence*. Portland, OR: American Association of Artificial Intelligence, 1996, pp. 725–730.

Quinlan JR, Cameron-Jones RM. Induction of logic programs: FOIL and related systems. *New Generation Comput.* 1995; 13: 287–312.

Quinn JF. On the statistical detection of cycles in extinctions in the marine fossil record. *Paleobiology* 1987; 13: 465–478.

Raz J, Zheng H, Ombao H, Turetsky B. Statistical tests for fMRI based on experimental randomization. *Neuroimage* 2003; 19: 226–232.

Reiczigel, J. Bootstrap tests in correspondence analysis. *Appl. Stochastic Models Data Anal.* 1996; 12: 107–117.

Reynolds RG, Ali M, Jayyousi T. Mining the social fabric of archaic urban centers with cultural algorithms. *Computer* 2008; 41: 64–72.

Ringrose TJ. Bootstrapping and correspondence analysis in archaeology. *J. Archaeol. Sci.* 1992; 19: 615–629.

Ritland C, Ritland K. Variation of sex allocation among eight taxa of the *Minimuls guttatus* species complex (Scrophulariaceae). *Am. J. Botany* 1989; 76.

Robeson SM. Resampling of network-induced variability in estimates of terrestrial air temperature change. *Climate Change* 1995; 29: 213–229.

Romano JP. A bootstrap revival of some nonparametric distance tests. *JASA* 1988; 83: 698–708.

Romano JP, Shaikh, AM, Wolf M. Control of the false discovery rate under dependence using the bootstrap and subsampling (with discussion). *Test* 2008; 17: 417–442.

Romney AK, Moore CC, Batchelder WH, et al. Statistical methods for characterizing similarities and differences between semantic structures. *Proc. Natl. Acad. Sci. USA* 2000; 97: 518–523.

Roper RJ, Doerge RW, Call SB, Tung KSK, Hickey WF, Teuscher C. Autoimmune orchitis epididymitis and vasitis are immunogenetically distinct lesions. *Amer. J. Path.* 1998; 152: 1337–1345.

Rosenbaum PR. Permutation tests for matched pairs with adjustments for covariates. *Appl. Stat.* 1988; 37: 401–411.

Roy T. Bootstrap accuracy for nonlinear regression models. *J. Chemometrics* 1994; 8: 37–44.

Royaltey HH, Astrachen E, Sokal RR. Tests for patterns in geographic variation. *Geogr. Anal.* 1975; 7: 369–395.

Ryan JM, Tracey TJG, Rounds J. Generalizability of Holland's structure of vocational interests across ethnicity, gender, and socioeconomic status. *J. Counseling Psychol.* 1996; 43: 330–337.

Ryan TP. *Modern regression methods.* New York: John Wiley & Sons, 1997.

Salmaso L. Synchronized permutation tests in 2^k factorial designs. *Commun. Statist. Theory and Methods.* 2003; 32: 1419–1438.

Sandford BP, Smith SG. Estimation of smolt-to-adult return percentages for Snake River Basin anadromous salmonids. *J. Agric. Biol. Environ. Stat.* 2002; 7: 243–263.

Santner TJ, Snell MK. Small-sample confidence intervals for p_1–p_2 and p_1/p_2 in 2×2 contingency tables. *JASA* 1980; 75: 386–394.

Schall R. Assessment of individual and population bioequivalence using the probability that bioavailabilities are similar. *Biometrics* 1995; 51: 615–626.

Scheffe H. *Analysis of variance.* New York: John Wiley & Sons, 1959.

Schenker N. Qualms about bootstrap confidence intervals. *JASA* 1985; 80: 360–361.

Sen B, Banerjee M, Woodroofe M, Mateo M, Walker M. Streaming motion in Leo I. *Ann. Appl. Stat.* 2009; 3: 96–116.

Senar JC, Conroy MJ. Multi-state analysis of the impacts of avian pox on a population of Serins (*Serinus serinus*): the importance of estimating recapture rates. *Animal Biodiversity Conservation* 2004; 27: 133–146.

Shao J, Tu D. *The jackknife and the bootstrap.* New York: Springer, 1995.

Shen CD, Quade D. A randomization test for a three-period three-treatment crossover experiment. *Commun. Stat. B* 1986; 12: 183–199.

Shimbukaro FI, Lazar S, Dyson HB, Chernick MR. A quasi-optical method for measuring the complex permittivity of materials. *IEEE Trans. Microwave Theor. Technol.* 1984; 32: 659–665.

Shuster JJ. *Practical handbook of sample size guidelines for clinical trials.* Boca Raton, FL: CRC Press, 1993.

Shuster JJ, Boyett JM. Nonparametric multiple comparison procedures. *JASA* 1979; 74: 379–82.

Siemiatycki J. Mantel's space-time clustering statistic: computing higher moments and a comparison of various data transforms. *J Stat. Comput. Simul.* 1978; 7: 13–31.

Siemiatycki J, McDonald AD. Neural tube defects in Quebec: a search for evidence of 'clustering' in time and space. *Br. J. Prev. Soc. Med.* 1972; 26: 10–14.

Simon JL. *Basic research methods in social science.* New York: Random House, 1969.

Simmons RB, Weller SJ. What kind of signals do mimetic tiger moths send? A phylogenetic test of wasp mimicry systems (Lepidoptera: Arctiidae: Euchromiini). *Proc. R. Soc. Lond. B. Biol. Sci.* 2002; 26: 983–990.

Smith PWF, Forester JJ, McDonald JW. Monte Carlo exact tests for square contingency tables. *J. R. Stat. Soc. A* 1996; 59: 309–21.

Smythe RT. Conditional inference for restricted randomization designs. *Ann. Math. Stat.* 1988; 16: 1155–1161.

Sohn I, Owzar K, George SL, Kim S, Jung SH. A permutation-based multiple testing method for time-course microarray experiments. *BMC Bioinformatics.* 2009; 10: 336.

Solomon H. Confidence intervals in legal settings. In *Statistics and the Law.* De Groot MH, Fineberg SE, and Kadane JB, ed. New York: John Wiley and Sons; 1986 455–473.

Somerville PN, Hemmelmann C. Step-up and step-down procedures controlling the number and proportion of false positives. *Comput. Stat. Data Anal.* 2008; 52: 1323–1334.

Steadman HJ, Silver E, Monahan J, Apelbaum PS, Robbins PC, Mulvey EP, Grisso T, Roth LH, Banks S. A classification tree approach to the development of actuarial violence risk assessment tools. *Law Hum. Behav.* 2000; 24: 83–100.

Stewart C, Tsai CL, Roysam B. The dual-bootstrap iterative closest point algorithm with application to retinal image registration. *IEEE Trans. Med. Imaging* 2003; 22: 1379–1394.

Streitberg B, Roehmel J. On tests that are uniformly more powerful than the Wilcoxon-Mann-Whitney test. *Biometrics* 1990; 46: 481–484.

Stuart GW, Maruff P, Currie J. Object-based visual-attention in luminance increment detection. *Neuropsychologia* 1997; 35: 843–853.

Subramanian A, Tamayoa P, Moothaa VK, Mukherjeed S, Eberta BL, Gillettea MA, Paulovich A, Pomeroy SL, Goluba TR, Landera ES, Mesirova JP. Gene set enrichment analysis: a knowledge based approach for interpreting genome-wide expression profiles. *Proc. Natl. Acad. Sci. USA* 2005; 102: 15545–15550.

Sukhatme BV. A two-sample distribution-free test for comparing variances. *Biometrika* 1958; 45: 544–548.

Svetnik V, Liaw A, Tong C, Wang T. *Application of Breiman's random forest to modeling structure-activity relationships of pharmaceutical molecules.* Berlin: Springer, 2004.

Tambour M, Zethraeus N. Bootstrap confidence intervals for cost-effectiveness ratios: some simulation results. *Health Econ.* 1998; 7: 143–147.

Tan Q, Brusgaarda K, Torben A, Krusea TA, Edward Oakeley E, Brian Hemmings B, Henning Beck-Nielsena H, Hansenc L, Gastera M. Correspondence analysis of microarray time-course data in case-control design. *J. Biomed. Informatics* 2004; 37: 358–365.

Tang L, Duan N, Klap R, Rosenbaum-Asarnow J, Belin TR. Applying permutation tests with adjustment for covariates and attrition weights to randomized trials of health-services interventions. *Stat. Med.* 2009; 28: 65–74.

Tian L, Greenberg SA, Kong SW, Altschuler J, Kohane IS, and Park J. Discovering statistically significant pathways in expression profiling studies. *PNAS. USA.* 2005; 102: 13544–13549.

Tibshirani RJ. Variance stabilization and the bootstrap. *Biometrika* 1988; 75: 433–444.

Tittonell P, Shepherd KD, Vanlauwea B, Giller KE. Unravelling the effects of soil and crop management on maize productivity in smallholder agricultural systems of western Kenya—an application of classification and regression tree analysis. *Agric. Ecosyst. Environ.* 2008; 123: 137–150.

Tivang JG, Nienhuis J, Smith OS. Estimation of sampling variance of molecular marker data using the bootstrap procedure. *Theor. Appl. Genet.* 1994; 89: 259–264.

Townsend RL, Sklski JR, Dillingham P, Steig TW. Correcting bias in survival estimation resulting from tag failure in acoustic and radiotelemetry studies. *J. Agric. Biol. Environ. Stat.* 2006; 11: 183–196.

Tracy DS, Khan KA. Comparison of some MRPP and standard rank tests for three equal sized samples. *Commun. Stat. B* 1990; 19: 315–333.

Tracy DS, Tajuddin IH. Empirical power comparisons of two MRPP rank tests. *Commun. Stat. A* 1986; 15: 551–570.

Trieschman JS and Pinches GE. A multivariate model for predicting financially distressed P-L insurers. *J. Risk Insurance.* 1973; 40: 327–338.

Troendle JF. A stepwise resampling method of multiple hypothesis testing. *JASA* 1995; 90: 370–378.

Trujillano J, Badia1 M, Serviá1 L, March J, Rodriguez-Pozo A. Stratification of the severity of critically ill patients with classification trees. *BMC Med. Res. Method.* 2009; 9: 83.

Tseng CC. Comparing artificial intelligence systems for stock portfolio selection. In *Proceedings of the 9th International Conference of Computing in Economics and Finance*, July 11–13 2003. Seattle; University of Washington.

Tsuji R. The structuring of trust relations in groups and the transition of within-group order. *Sociol. Theor. Method* 2000; 15: 197–208.

Tsutakawa RK, Yang SL. Permutation tests applied to antibiotic drug resistance. *JASA* 1974; 69: 87–92.

Tu D, Zhang L. Jackknife approximations for some nonparametreic confidence intervals of functional parameters based on normalizing transformations. *Comput. Stat.* 1992; 7: 3–5.

Tubert-Bitter P, Letierce A, Bloch DA, Kramer A. A nonparametric comparison of the effectiveness of treatments. A multivariate toxicity-penalized approach. *J. Biopharm. Stat.* 2005; 15: 129–142.

Tukey JW. Improving crucial randomized experiments—especially in weather modification—by double randomization and rank combination. In *Proceedings of the Berkeley Conference in Honor of J. Neyman and J. Kiefer*, LeCam L, Binckly P, eds. Vol. 1. Hayward, CA: Wadsworth, 1985, pp. 79–108.

Tukey JW, Brillinger DR, Jones LV. *Management of weather resources: The role of statistics in weather resources management.* Vol II. Washington, DC: Department of Commerce, U.S. Government Printing Office, 1978.

Valdes-Perez RE. Some recent human-computer studies in science and what accounts for them. *AI Mag.* 1995; 16: 37–44.

Valle D, Pesarin F, Salmaso L. Paired permutation testing in 2^k unreplicated factorials. *Stat. Meth. Appl.* 2002; 11: 265–276.

vanKeerberghen P, Vandenbosch C, Smeyers-Verbeke J, Massart DL. Some robust statistical procedures applied to the analysis of chemical data. *Chemometrics Intelligent Lab. Syst.* 1991; 12: 3–13.

Vanlier JB. Limitations of thermophilic anaerobic waste-water treatment and the consequences for process design. *Antonie Van Leeuwenhoek Int. J. Gen. Mol. Microbiol.* 1996; 69: 1–14.

van-Putten B. On the construction of multivariate permutation tests in the multivariate two-sample case. *Stat. Neerlandica* 1987; 41: 191–201.

Vickers AV, Cronin AM, Elkin EB, Gonen M. Extensions to decision curve analysis, a novel method for evaluating diagnostic tests, prediction models and molecular markers. *BMC Med. Informatics Decision Making* 2008; 8: 53.

Wald A, Wolfowitz J. Statistical tests based on permutations of the observations. *Ann. Math. Stat.* 1944; 15: 358–372.

Wei LJ. Exact two-sample permutation tests based on the randomized play-the-winner rule. *Biometrika* 1988; 75: 603–605.

Wei LJ, Smythe RT, Smith RL. K-treatment comparisons with restricted randomization rules in clinical trials. *Ann. Stat.* 1986; 14: 265–274.

Welch WJ. Rerandomizing the median in matched-pairs designs. *Biometrika* 1987; 74: 609–614.

Welch WJ. Construction of permutation tests. *J. Am. Stat.* 1990; 85: 693–698.

Welch WJ, Guitierrez LG. Robust permutation tests for matched pairs designs. *JASA* 1988; 83: 450–61.

Westfall DH, Young SS. *Resampling-based multiple testing: examples and methods for p-value adjustment*. New York: John Wiley, 1993.

Weth F, Nadler W, Korsching S. Nested expression domains for odorant receptors in zebrafish olfactory epithelium. *Proc. Natl. Acad. Sci. USA* 1996; 93: 13321–1332.

Whaley FS. The equivalence of three individually derived permutation procedures for testing the homogenity of multidimensional samples. *Biometrics* 1983; 39: 741–745.

Wheldon MC, Anderson MJ, Johnson BW. Identifying treatment effects in multi-channel measurements in electroencephalographic studies: multivariate permutation tests and multiple comparisons. *Austr. New Zeal. J. Stat.* 2007; 49: 397–413.

Wilk MB. The randomization analysis of a generalized randomized block design. *Biometrika* 1955; 42: 70–79.

Wilk MB, Kempthorne O. Some aspects of the analysis of factorial experiments in a completely randomized design. *Ann. Math. Stat.* 1956; 27: 950–984.

Wilk MB, Kempthorne O. Nonadditivities in a Latin square design. *JASA* 1957; 52: 218–236.

Witztum D, Rips E, Rosenberg Y. Equidistant letter sequences in the Book of Genesis. *Stat. Sci.* 1994; 89: 768–76.

Wong MA, Lane T. A kth nearest neighbor clustering procedure. *JRSS B* 1983; 45: 362–368.

Wu CFJ. Jackknife, bootstrap, and other resampling methods in regression analysis (with discussion). *Ann. Stat.* 1986; 14: 1261–1350.

Wu JC, Bell K, Najafi A, Widmark C, Keator D, Tang C, Klein E, Bunney BG, Fallon J, Bunney WE. Decreasing striatal 6-fdopa uptake with increasing duration cocaine withdrawal. *Neuropsychopharmacology* 1997; 17: 402–409.

Yucesan E. Randomization tests for initialization bias in simulation output. *Naval Res. Logistics.* 1993; 40: 643–663.

Zempo N, Kayama N, Kenagy RD, Lea HJ, Clowes AW. Regulation of vascular smooth-muscle-cell migration and proliferation *in vitro* and in injured rat arteries by a synthetic matrix metalloprotinase inhibitor. *Art. Throm. V* 1996; 16: 28–33.

Zerbe GO. Randomization analysis of the completely randomized design extended to growth and response curves. *JASA* 1979a; 74: 215–221.

Zerbe GO. Randomization analysis of randomized block design extended to growth and response curves. *Commun. Stat. A* 1979b; 8: 191–205.

Zerbe GO, Murphy JR. On multiple comparisons in the randomization analysis of growth and response curves. *Biometrics* 1986; 42: 795–804.

Zerbe GO, Walker SH. A randomization test for comparison of groups of growth curves with different polynomial design matrices. *Biometrics* 1977; 33: 653–657.

Zhang K, Zhao H. Assessing reliability of gene clusters from gene expression data. *Funct. Integr. Genomics* 2000; 1: 156–173.

Zoubir AM, Boashash B. The bootstrap and its application in signal processing. *IEEE Signal Process. Mag.* 1998; 15: 56–76.

Zucker S, Mazeh T. On the statistical significance of the Hipparcos astronomic orbit of ρ Coronae Borealis. http://arxiv.org/abs/astro-ph/0104098

Index